T0230739

Media Relations in Property

Graham Norwood & Kim Tasso

2006

Routledge
Taylor & Francis Group

LONDON AND NEW YORK

First published 2006 by Estates Gazette

2 Park Square, Milton Park, Abingdon, Oxon OX14 4RN
711 Third Avenue, New York, NY 10017, USA

Routledge is an imprint of the Taylor & Francis Group, an informa business

First issued in hardback 2017

ISBN 978-0-7282-0491-1 (pbk)
ISBN 978-1-138-46133-8 (hbk)

Typeset in Palatino 10/12

Contents

1 Media Relations — Why is it Important? 1
What is media relations? Corporate versus product media relations. Why is media relations important? Why and what to invest in media relations? Residential property publicity opportunities. Commercial property publicity opportunities.

2 PR Structures for Property Organisations 29
Why integrate media relations into your marketing? Skills for effective media relations. Knowledge and experience required. Appointing an in-house PR officer. Appointing an external PR agency. Costs. Supervising and managing your PR resources. Common problems and how to resolve them.

3 What Journalists Want and How You Can Provide It 61
The editorial market. Pitching commercial property stories. Press releases: Good and bad examples. Good pictures are vital. The importance of case studies. Exclusives. Means of communication. Learning the lessons.

4 Reactive PR: Direct Contact with Journalists 93
Reasons why journalists may contact you direct. Handling enquiries from national media property journalists. Handling interviews with friendly journalists on local papers.

Handling interviews with unfriendly journalists on national or local media. Handling enquiries on a negative story. Handling local or national television. Managing radio interviews. Direct contact with journalists via e-mail. Multiple outlets for an interview. Handling journalists on press trips. Don't panic.

Raise awareness of the need for media relations. Education and training. Communicate internally. Develop a media relations plan. Formulate policies and procedures. Prepare press releases. Distribute press releases. Organise press launches and conferences. Use research. Enter awards. Have lunch with journalists. Have a journalist to lunch. Corporate events. Attend exhibitions. Create "rock stars". Write feature articles. Other approaches. Building relationships with journalists.

Why measure media relations effectiveness? Difficulties in measuring media relations effectiveness. Measuring media relations effectiveness. Setting clear goals. Measuring the process. Measuring the results.

National media contacts. Property and construction media contacts. Broadcast media contacts.

Acknowledgements

The authors would like to thank the many people who contributed to and helped with this book, in particular:

Lucia Adams of *The Times*; Area Sq; Atisreal; Richard Barraclough of Try Homes; Barratt Howe Public Relations; Melanie Bien of Savills Private Finance; Carter Jonas; Chase & Partners; City Lofts Ltd; Clucas Communications; Company of Chartered Surveyors; Margie Coldrey of MCPR; Alison Dean of JagoDean; Jeremy Dodd, PR consultant; Zoe Dare Hall, journalist; Margaret Emmens, Olivia Smith and Emma Stanley-Evans of Knight Frank; English Partnerships; First Counsel; Paul Gray of King Sturge; Matt Havercroft of *A Place in the Sun*; haysmacintyre; Roger Hunt, journalist; Jill Knight, journalist; Terry Knight; Nigel Lewis of the *Daily Mail*; Madeleine Lim of *The Independent*; Cheryl Markosky, journalist; Andy Martin of Harrison Cowley PR; Mattison PR; Joy Moon of Strutt & Parker; Debra Morrall of Iris PR; Catherine Moye, journalist; Nightingales PR; Pellings; PROFILE; PSMG; Purple Cake PR; Alison Richards and Adam Tinworth of *Estates Gazette*; RICS; Rok property solutions; Russell Ross-Smith of Hamptons International; Bertie Sanderson of Primmo PR; Sapcote Developments; Schillings; Jon Shilling, formerly of Golder Associates; Carey Scott of the *Sunday Times*; SJBerwin; Slough Estates; Phil Spencer of Garrington Home Finders; Tim Stanley, PR consultant; Christopher Stoakes; Giles Taylor of Marketing Resources; Taylor Woodrow; TRMA public relations; John Vaughan of Savills; Nicola Venning, journalist; Mary Wilson, journalist.

"For Helen, with thanks for her endless patience."
Graham Norwood

"For my children — James and Lizzie."
Kim Tasso

Foreword

This is an alarmingly good book. Good, because it sets out with concision, clarity and great accuracy the challenges facing those who deal with the press. For journalists the book is quite alarming. That is because the authors have clearly got the measure of the national, trade, and regional property press! Those dealing with the promotion of residential sales will find the book just as useful as those whose job it is to further the aims of the clients who deal in the commercial sector.

Journalists like to think they can manage without PR people. PR people like to think they can manage journalists. The truth lies somewhere in between. Those seeking the truth can do no better to read this well written, highly accurate and entertaining guide to the sometimes fraught relationships between the property media and those seeking to influence what gets printed.

Peter Bill

Preface

PR gets a bad press.

Hardly a day goes by without someone having a go at the warped and dishonest world of "spin doctors". Bolstering the egos of talentless pop stars and supporting the flagging careers of slippery politicians is a lowly occupation, a million miles from the quiet and careful planning of the property developer or the busy reality of a residential agent and the insightful research of the well-intentioned surveyor.

So has PR lost its credibility? Has PR fallen from the grace of a respectable trade to a dark world where each word is interpreted as a lie and difficult items become "buried" during fast news days? And why on earth should the busy property professional get involved?

Good PR, as with good salesmanship, is hardly noticed. It achieves its goals and everyone is happy. The newspaper or magazine gets a great story and the company or individual generates useful publicity. Free copy for the magazine and free advertising for the company — a win/win outcome.

Yet good PR, again like good salesmanship, takes skill, hard work, time, effort and patience. It takes research, time and perseverance to build relationships. This book aims to help you develop the skills and attributes of a great PR and to deploy them in a way that results in great publicity for your property endeavour.

More now than ever, we live in a media world.

We cannot afford to ignore such a powerful and sometimes dangerous communications channel and business tool. Professional PR and its enormous contribution to the success of countless companies, people, communities and individuals must not be lost in the shadows of bad spin. This book shows what good PR is, how to do it right and what it can do for you in the property industry.

We will demystify "the black art of PR". We will explain what it is and what it isn't, and what it can and cannot achieve. We will look at how it can help you whether you are a developer, a surveyor, an architect, an agent or someone on the periphery of the industry. We hope it will help whether you are in a large or small organisation, the commercial or residential market, already have professional PR support or are just a beginner.

We will explain the role of professional PR in a modern organisation's overall marketing and business development plans — even how it can support other aspects of business strategy. We will also consider the role of PR in developing the great brands of the property industry.

We will show how PR adds significant value and reduces cost on major sales campaigns. We will consider the business benefits — whether you are promoting your organisation, your services, your buildings, your people or yourself — and we will consider how PR can help you convey your expertise or persuade people around to your way of thinking.

But we will also manage your expectations.

PR is not magic and many people are disappointed with the results because they were unclear about their goals, unrealistic about what could be reasonably achieved, or because they used the wrong methods. Sometimes they failed to understand the constraints and aims of the journalists and media they target.

We have devoted significant space in this book to describing how the media works, the needs of different types of journalist and media and how you might work with them most effectively.

Rather ambitiously, we have tried to provide something for everyone:

- simple explanations
- insights into the way the media works
- case studies showing how others have achieved great publicity and some of the disappointments they have had
- pragmatic advice on how to deal with journalists
- whether to appoint in-house media professionals or hire an agency
- how to manage your PR professionals on an ongoing basis
- how to identify and communicate a truly good story
- how to generate positive PR for your organisation or development.

Finally, being ever conscious of the commercial nature of business, we also consider how much to invest and the various ways in which you can measure your return on PR investment.

How should you use this book?

Well, you could read it from start to finish — that is how we intended for it to be read. But busy senior executives might concentrate on the business case in chapter 1 and how to resource it in chapter 2 and measure its effectiveness in chapter 6. Marketing professionals who wish to gain a better understanding of PR and start to implement an approach should concentrate on the insights into journalists in chapter 3 and the various reactive and proactive techniques in chapters 4 and 5. Those in the PR industry might learn from the insights into the media in chapter 3.

You should keep it as a reference document and use the glossary and index to help you identify relevant advice on how to tackle particular issues if and when they arise.

As with all good journalism, the book attempts to present both sides of the story — the marketing / PR professional on the one hand and the seasoned journalist on the other. The authors bring together a range of skills and experience from these and other perspectives. But we also had plenty of help from many people in other sectors of the marketing, media and property world too. We hope you find it useful. And we hope you will send your comments, stories and ideas for the next edition.

The pen is mightier than the sword.

Graham Norwood and Kim Tasso

About the Authors

Graham Norwood is a freelance property writer, contributing each week to the *Sunday Times* and regularly appearing in *The Independent, The Observer, The Times,* the *Sunday Telegraph* and the *Financial Times.* He also writes on global property markets for international publications including the *Wall Street Journal.*

He appears in property lifestyle magazines including *A Place in the Sun* and is a frequent contributor on the residential industry and new technology issues for the leading property weekly *Estates Gazette.* He also provides media training to many leading estate agents and developers and is a judge at leading industry awards events such as the Estate Agency of the Year competition.

Graham's first book, *21st Century Estate Agency,* looks at the use of technology in the residential property industry in the UK and throughout the world, and was published by EG Books in 2005.

Until 2000, Graham was a journalist at the BBC. He now lives in Devon and details of his work can be found on *www.grahamnorwood.info.*

Kim Tasso BA(Hons) DipM MCIM MIDM MCIJ MBA is a strategic marketing and business development consultant. Property sector clients include: Allsop & Co, Atisreal, AYH plc, Crispin & Borst, IVSC, Pellings and Slough Estates. She has also worked with hundreds of legal, accounting and consultancy businesses — many of whom have significant involvement with the property sector.

As a freelance journalist, she has written articles on marketing and client development for magazines including: *Estates Gazette, Construction Marketer, Solicitors' Journal* and *Professional Marketing.*

She has a degree in psychology, a diploma in marketing, a diploma in professional coach/mentoring and an MBA.

The first seven years of her career were spent in marketing and selling roles in the technology sector, and one of her roles was running the press office for an American corporation. She went on to hold senior marketing positions at several leading professional service firms including Deloittes, Nabarro Nathanson and Atisreal. She also lectured part-time in post-graduate marketing courses. She is a regular trainer on a variety of marketing, business development and personal effectiveness courses and a judge on two leading marketing awards.

Her first book *Dynamic Practice Development — Selling skills and techniques for the professions*, published in June 2000, was revised and reprinted in June 2003.

She lives in Twickenham with her two children. Further information is available on *www.kimtasso.com*.

Media Relations — Why is it Important?

What is media relations?

We have to start with some definitions. We know you want to plunge into what to do and how to do it, but it is important we understand exactly what we are talking about when we consider media relations.

Many have observed that while the property sector is necessarily driven by the deal, there are too few who understand and apply the theory and application of good marketing and PR principles.

Marketing is a big subject. PR is too. We need to have a clear view of what PR and media relations are in order to understand where media relations fits into the grand scheme of things — how it fits in with the other promotional activities you might come across.

Take the term "PR" for instance. PR can stand for the term "Public Relations" and this itself encompasses a diverse range of activities including:

- Investor relations (talking to the stock markets, shareholders, investors, analysts etc).
- Internal communications (talking to your staff).
- Community relations (dealing with pressure groups and stakeholders in the community).
- Parliamentary affairs (lobbying).
- Client and customer relations.
- Recruitment marketing (the rise of "the employer brand").
- Media relations — dealing with the print, broadcast and electronic media that serve your clients, customers, tenants and your industry.

Definition

Public relations:

> Building up and keeping good relations between an organisation and the public, or an organisation and its employees, so that people know and think well of what the organisation is doing.

This book is about how your organisation communicates with one, rather special "public" — the media. And therein lies the confusion: sometimes PR is used to describe press relations, which, if you extend to include broadcast and electronic media, is synonymous with media relations, hence the title of this book.

So media relations is one aspect of public relations. And public relations is, in turn, part of marketing communications, which is part of the overall marketing mix.

Marketing is the overall process designed to consider your marketplaces and products, prices and people and then to promote or communicate the key messages effectively. Your PR or media relations strategy is unlikely to be effective unless it is an integrated part of the overall marketing strategy which, in turns, translates the needs of the business plan into a series of marketing objectives and programmes.

Definition

Marketing:

> The management process responsible for anticipating and meeting customers' needs profitably.

Some describe marketing as "Getting the right product (or service) at the right price to the right place at the right time with the right promotion". This definition of marketing is rarely understood in the property industry. Most people equate marketing with just the marketing communications element of the marketing mix: advertising, brochures, websites and parties. Too often, the important role of market research, market planning, product (service) development, pricing and staff training and communications are overlooked. Too often, marketing does not take place at the heart of the property business — where the market trends and needs shape every aspect of an organisation's strategy — but at the sharp end of promotion and selling alone.

Table 1.1 The marketing mix and how PR (public relations) fits in

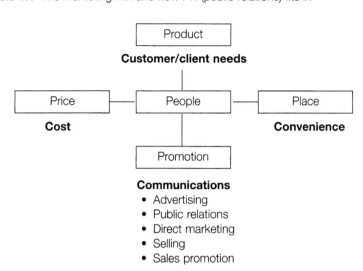

But promotion, or marketing communications as it is sometimes called, is just one aspect of the marketing mix. Although it is the most apparent and easiest to recognise, it should only follow once all the hard work is done on working out the market needs, the impact of economic trends and competitors' actions, investment in producing the right product or service that is sufficiently differentiated or available at a special price and efforts to ensure that everyone within the organisation understands the main proposition and messages and their role in delivery. Often in the property industry, these vital marketing activities are undertaken by the senior management and divorced from the marketing department within a property business.

Marketing communications uses five main types of tools — many of which are seen as being deployed effectively in both the commercial and residential property industries on a regular basis.

Advertising — This is where you pay a media owner (a magazine, a poster site owner, a radio station) for your message to appear in exactly the way you define. The advertisement design needs to grab attention. The advertising message needs to be simple. Your advertisement needs to prompt some form of action that can be measured. Advertising in national newspapers, trade magazines, on poster sites, on transport systems or on hoardings can be expensive. And your advert must fight against all the other advertising messages that are vying for the attention of your target audience.

Advertising is one of the main approaches used by those targeting large groups of consumers — for example, in the residential property market. It is a key tool in helping an organisation to build a high profile brand. But it is also used in the commercial market — for example, by drawing the attention of industrial agents to the availability of newly available industrial warehousing schemes through the specialist property media. On a simpler and more cost-effective level, the use of a good letting board or hoardings act as an excellent advertisement to those in the immediate area around a new development.

It is probably worth highlighting at this point a vital difference between media relations and advertising, as they are sometimes confused.

Media relations is where we communicate with journalists and provide them with information that encourages them to write about us. There are no guarantees of what — if anything — will happen to the information you provide. With advertising, you are dealing with, usually, a completely different set of people who spend their time selling space. These differences are explored further below. And you pay for that space and get your designers or advertising agents to produce an advert that then appears, exactly as you wish, among the other advertisements.

Direct marketing — Direct mail is used by both commercial and residential approaches. It requires the purchase or development of an appropriate list, the production of an attention-grabbing mail piece and a suitably exciting offer to promote action. It can be costly and slow to manage all the printing and postage, although the advent of e-mail has reduced costs and made the whole process quicker. The use of electronic media for direct marketing has also enabled tighter targeting. Yet it has also brought with it new legislation to protect the use of people's information and their privacy.

Although cheaper than advertising, it has less reach, it enables you to target your message more precisely to those individuals and businesses you are most interested in reaching. Telemarketing — where telephone calls are made to targets — is carefully regulated now but enables a dialogue to be established.

Sales promotion — These are usually short-term, price-related mechanisms designed to generate fast action. Residential developers often offer free legal fees or mortgage assistance to encourage first-time buyers to take the plunge. Commercial developers will offer inducements and prizes to persuade agents to visit their sites and rent-free periods and other inducements to business tenants to take leases. Quite often, both residential and commercial developers might include attractive giveaways in direct mail campaigns to encourage people to visit their development or arrange a viewing.

Selling — Once the other methods of marketing have captured the attention of your audience, and prompted them to make contact, you deploy sales techniques to move them closer to a sale. Telephonists in call centres, staff managing viewings in show homes and agents who take enquiries are all involved in selling in order to match the needs of the customer against the products or services they have to offer. With the right people it is highly effective, but also it is very expensive. The majority of letting agents, whether in-house or in private practice, devote a significant amount of their time to selling, particularly in the commercial market where there are usually many people (including legal advisers) involved in the selling and negotiating process.

Public relations — This is a catch-all category for a wide variety of other marketing activities designed to convey an organisation's messages to its publics. It uses a host of activities to support the other marketing techniques including publications, events, exhibitions, websites, sponsorship and corporate hospitality. And, of course, the all-important activities to support good media relations.

Table 1.2 The elements of PR (public relations) in the property sector

PR specialism	Target audience
• Parliamentary affairs	• Government
• Investor relations	• Financial markets
• Community relations	• The community
• Internal communications	• Your staff
• Recruitment consultancy	• Your potential staff
• Consumer relations	• Consumers
• B2B/client relations	• Commercial organisations
• Referrer management	• Intermediaries
• Media relations	• The media

But why are all these definitions and explanations important to you?

If you have a marketing challenge, then you need to know what the appropriate tool might be — advertising works in a very different way from media relations — and the skills, costs, methods, timescales and outcomes are different too.

You also need to appreciate the different marketing activities in order to understand how they can be used together in an integrated way, reinforcing messages at different stages of the buying or selling process. It is important to see how media relations might work in conjunction with other marketing activities to achieve an overall impact that is much greater than the results achieved with just one particular activity.

If you are hiring in-house staff or appointing an external agency, and planning to spend a significant amount to achieve your business aims, you need to know what you are hiring. There are many instances of property companies spending significant amounts on some marketing or PR consultant only to learn after many tears and considerable cash that they are using the wrong type of adviser to try to achieve their aims.

A PR consultant can be a very different animal to a media relations consultant, and completely different from a marketing adviser.

Public relations consultants will typically have a wide range of skills and be able to do a great job for you in many areas — they will help you produce brochures and publications that explain your messages clearly. They will design and manage fantastic events to get people together in a setting suitable for you to do business. They will project manage your attendance at major trade shows and exhibitions. They will make your people enthusiastic about your products and services. They will organise sponsorships that raise your profile, add cachet to your senior people and provide excellent hospitality opportunities.

Of course, they will all be aware of the different activities but, as you find surveyors who specialise in rating, valuations, agency, finance or development, you need to get the right expert for the job in hand. Some PR consultancies also provide a media relations service. But make sure that the necessary media relations skills and contacts are held by those working on your account.

So, when we talk about media relations we are talking about the specific activities surrounding how your organisation communicates effectively with the media, be that the national press, property trade magazines, local newspapers, specialist media reaching your residential buyers and tenants or your commercial investors and occupiers. In addition, you must consider the broadcast media —

television, radio, the internet and its numerous property portals and websites, as well as electronic magazines and newspapers.

Corporate versus product media relations

In the property sector, as in several others, there is an interesting divide within media relations.

Corporate PR

On the one hand, there is what is often called "corporate PR" where the aim is to convey positive messages to the media regarding your organisation, its services and its people. Sometimes corporate PR is likened to reputation management. It is about your organisation, its goals, values, strategies and senior people. Corporate PR is particularly important to the professions and service businesses within the property industry — it enables an organisation to retain a high profile, to be seen to be acting on and contributing to a variety of debates, for its staff to be respected for their opinions and expertise.

These days, corporate social responsibility (CSR) is an important issue for investors, customers, staff and the community at large — in fact, all stakeholders in a company. Corporate PR is a vital tool in communicating your organisation's mission and values in relation to CSR.

Corporate media relations enables an organisation to convey the right messages to the right media to achieve the appropriate reputation in a low-key and credible way. Corporate PR is about building the corporate brand, the core values with which the organisation wishes to be associated. This is sometimes called "the employer brand".

But for these reasons corporate media relations can be difficult — there is often no product to show in a photo, perhaps only pictures of people associated with their opinions and comments. Apart from the annual results, which has a relatively high news value for larger or public companies, it takes hard work and imagination to ensure that positive and relevant information flows from the organisation to the media during the rest of the year.

There is also much work to do to ensure that any negative or difficult situations are handled appropriately, either minimising the media impact of negative coverage or ensuring that your organisation's side of the story is presented fairly. Corporate PR requires a sustained effort on an ongoing basis — the need to work at it never stops — and therefore it is reliant on close relationships and regular contact with those media that are most likely to cover stories about your organisation.

Product PR

"Product PR" is different. It is much more focused and is where media relations is deployed for a specific period of time in a campaign to achieve some specific short-term goal.

It is usually focused on something far more tangible than an organisation. It might be related to the promotion of a particular product — a development, a commercial building or a residential property. Residential developers may have a few short months to raise the profile of a new development to secure a high take-up to generate cash for the business. For service providers, it may be concerned with highlighting the availability of, for example, specialist rating expertise at a time when many commercial occupiers are facing a revaluation.

Product PR is usually concerned with raising the profile of the product within a geographic area or a segment of the consumer or commercial market. This will mean the range of media targeted may be very different from that targeted with corporate PR. Product PR will sometimes have the slightly longer-term task of building a brand for a particular location or development, although this is usually easier to define and communicate than the more intangible environment of a corporate brand.

So while the principles are similar, there are four big differences.

Time frame and sustainability — Corporate media relations needs to be an ongoing and consistent effort. There needs to be a strong link between the senior management of the organisation and the media relations function to ensure there are strong relationships with the key journalists and a constant flow of good material that can be used by the media. It is, in marketing terms, a drip-drip effect. Product PR is usually confined to a shorter time frame and needs a more concentrated burst of activity to coincide with key stages in the development — groundbreaking, completion, viewing, letting, selling etc. The media relations campaign lasts for a short while and is more transactional.

Nature of media — Corporate media relations usually focuses on the national media (for large organisations) and the property trade magazines. There will also be a myriad of other business-to-business media serving the particular target markets and client groups served by the organisation. In product media relations, the target media will often be within consumer magazines and newspapers and perhaps also the regional and local media read by consumers. Product PR is far more likely to have to become involved with local television and radio stations as the number of people — particularly in residential campaigns — is likely to be broader and more dispersed.

Responsibility and payment — While corporate PR is usually managed by someone within the senior management of the firm, a marketing director or partner, product media relations is usually managed by the property professionals and their advisers with responsibility for the development or property. Often the bill is paid by the owner of the building or development or perhaps the managing agent. This often means that the product media relations is not managed by a marketing or PR professional — and may mean that conflicting messages about the organisation from corporate and product media campaigns reach the same media.

Results — It is notoriously hard to measure the effectiveness of corporate PR. How do you measure the extent to which your organisation is viewed positively by all those different audiences? How do you assess whether your corporate PR effort is generating the same, more, or less publicity than last year? Product PR is a little easier to measure — there will be enquiries generated as a result of the campaign, hopefully an increase in web traffic and the commercial units or residential homes will be sold or let on time and at the right price.

Why is media relations important?

This could best be answered by those inside the property industry rather than by journalists or media relations experts.

One persuasive reason comes from Terry Knight, Master of the Company of Chartered Surveyors in 2004/05.

He says:

Sadly, the property industry is still viewed in some areas with some disdain; associated with Rachmanites in the residential arena and fat cats in the commercial market. Too many people — in the commercial world as well as the general public — don't realise the value of property as the day-to-day impact on the built environment, the huge commercial investment market opportunities and as the bedrock of many millions of people's pensions.

There is a continued need to educate people in property's role and value and to raise the profile of the property industry generally and good media relations is the tool we must use to ensure that accurate information and a true picture triumphs over sensationalism. Bad news makes good copy — so the trick is to make good news interesting and relevant.

So how do we achieve Terry's aims?

A major part of media relations is focused on getting products and corporate brands mentioned in the large number of print, online and broadcast media available today, with good reason. As we shall see, written or verbal editorial comment is not only less expensive than simple advertising, but also it is deemed more authoritative by consumers.

We need some more definitions to show media relations in action from the opposite side, from the journalists who are on the receiving end of much of media relations' output.

A key distinction is between advertising, which has already been described and defined, and "editorial".

Editorial is just about anything written in a publication or spoken broadcast that is not an advertisement. All editorial — whether "hard" news, sport, business or the more feature-oriented material such as property, travel or motoring — is written by journalists and should, in theory, be free of favouritism or bias.

If it has an angle, this will be done to reflect the typical readership or viewing demographic of the publication or broadcast but should still be neutral. So, for example, if a story is written in *The Sun* about the housing market it will major on average prices and the "volume" end of the market; if written for the *Daily Telegraph*, it may concentrate more on upper-end properties.

But whichever the audience, the article should be based on factual evidence and not unfairly talking the market up or down for no reason. Therefore, all editorial should present factual evidence to support a claim whether about figures on mortgage lending, the merits of a specific property on sale, or a suggestion that a celebrity is attempting to buy a country estate in Scotland.

Advertising (as we have seen, that element of a newspaper, magazine, website or broadcast that is paid for by individuals, companies or groups) can occasionally be a source of confusion if an advertiser deliberately uses "look-alike" ads to carry similar but not-quite-identical pages of information that may, at first glance, seem editorial. Quite often the content of these "advertorials" is not written by journalists.

For newspapers and most magazines, and for most broadcasters in the UK, the editorial content of their publication or programme is sacrosanct and is not influenced by advertising. If this suggestion causes a snigger among hard-nosed estate agents, developers, architects or surveyors, it should not.

Just ask London estate agency Foxtons precisely how much it spent advertising with Rupert Murdoch's News International organisation in summer 2003 while at exactly the same time some of Murdoch's best known news outlets — chiefly *The Times* and *Sky News* — were making allegations about the behaviour of some employees and associates of Foxtons.

Or ask the press office of commercial and residential agency Savills about the weeks in summer 2001 when two newspapers' property sections deliberately stopped including the firm's properties and spokesmen in editorial. This was done after Savills allegedly upset the two papers in an attempt to offer each publication an "exclusive" on the same much-prized country house that was for sale. At the same time, Savills was one of the largest advertisers in each newspapers' property section — a fact that made no difference to the temporary boycott of the firm by the papers in question.

There is one other piece of evidence that shows advertising and editorial to be quite separate. Many residential developers complain that they spend large sums in national newspaper advertising but that very few column inches are given to new homes, unlike the substantial coverage given to older properties or those with particular human interest stories associated with their sellers.

"If new homes didn't look the same and if more developers provided high-quality photographs instead of dreadful computer-generated shots of buildings with BMWs outside them on cloudless summer days, it would help" is the view of Madeleine Lim, editor of *The Independent*'s property section.

This may or may not be a justifiable view according to developers (and we will address the specific issue of photographs for the press, broadcast and electronic media later in this book) but Lim's comments are typical of most newspaper and magazine property editors. They explain why the mix of residential editorial is so unlike the mix of residential advertising in the press.

Other useful definitions at this stage concern the people at the sharp end of writing editorial, the property editors, staff and freelance property journalists, and sub-editors who pen the headlines, text and captions for papers, magazines, online and broadcast programmes.

- *Property editors* — these are the individuals who edit the property pages of a newspaper (say, the Home supplement of the *Sunday Times*) or the property pages of a magazine (for example, the property section of *Country Life*). These should not be confused with the overall editors of the publications.
- *Property journalists* — these are the journalists that most agents and developers come into contact with, and they fall into two categories.
  1. *Staff writers* (that is, employed by a newspaper or magazine, based in the publication's offices and writing solely for that title or series of titles). The editorial staff of *Estates Gazette* magazine fall into this category, as do the editor and senior journalists at some of the national newspaper property sections — the *Sunday Times* has four, the largest number of property editorial "staff writers" of any national paper.
  2. *Freelance property writers* (that is, journalists not employed by any one newspaper or magazine, but who write on commercial and/or residential property issues for a range of titles). The two authors of this book fall into this category, as do about 200 other journalists in the UK. Even publications with an extensive array of staff writers (*Estates Gazette*, for example) use freelances to top up the editorial content, while most national newspapers rely on freelances extensively.

There are more freelances than staff writers for financial reasons. Property sections in most magazines and newspapers are cash cows — for example, the London *Evening Standard*'s Wednesday supplement Homes & Property earns its owners, Associated Newspapers, more than any other supplement or section that it produces, even including the various pull outs and magazines with the *Daily Mail* and *Mail on Sunday*.

But if there were a slump in the housing market and advertising plummeted too, the reliance on freelances makes it easy for newspapers to cut back — they simply drop the use of freelances, which can be done at no notice and without any expensive redundancy payments.

Because freelances are only paid for stories that are "commissioned" by an editor and then published, they will only rarely work speculatively on a story before it is commissioned. A commission is an agreement that the story will be paid for when it is published, so long as it is up to standard and delivered on time. Therefore, freelances never waste time on stories that are not commissioned, in case they are never picked up by a publication.

Payments to freelances vary, and increasingly individual journalists strike their own rate with editors. As a guide, revelatory, celebrity-fixated property sections such as the *Mail on Sunday*'s will pay £800 for a 1000-word story if the journalist arranges case studies, an original story and ensures the subjects of the piece are willing to be photographed. That fee may hit £1000 if it is a lead story with pictures for the cover of the section, too. Heavyweight sections such as the *Sunday Times'* Home or the *Daily Telegraph* or *Financial Times'* property pages on Saturday pay £350–550, while *The Independent* or *The Observer* pay as little as £200–300.

The scale and status of exposure given to a story and the journalist who writes it means there is no shortage of contributors, even to the lower-paying nationals. But national magazines are a different problem — specialist trade publications such as *Estates Gazette* pay a modest £220 per 1000 words but has no real rival, while glossy overseas property monthlies tend to pay £300 per 1000. These often carry versions of stories previously carried in nationals.

Whether a staff or a freelance journalist, the typical property writer will want to have a continuing relationship with property professionals such as PR officers, estate agents, developers, architects, surveyors and designers.

This is not to say that a property journalist will be a soft touch because they rely on property professionals within the industry. To prove the point, professional journalists pride themselves on writing strident stories revealing the market is falling even if an agent says it is not, or stating that a development is not as perfect as its builder suggests.

But most property journalists have many years of experience, have dealt with all aspects of the property industry, and will want to speak to the same agents, developers and press relations officers (PROs) next week ... so will not do anything to unnecessarily antagonise them this week.

The next definition is that for *property sub-editors*, often unsung heroes in newspapers, whose work makes articles fit well on a page, read coherently, and not contradict or duplicate stories elsewhere. They will hardly ever speak directly with property professionals and will liaise mainly with the property editors and the property journalists. In simplistic terms, the subs' job consists of checking the accuracy of the content, making it fit the available space on the page or "spread" (that is, when a story runs across more than one page), writing captions for photographs, and writing a headline for the overall story.

The subs have the right to cut a journalist's work as he or she sees fit to suit the space, without recourse to the journalist. Sometimes, albeit rarely, the sub will add material if the original story filed is too short for the space available. Either way, it is common for a journalist to see his or her story in print or on a website, in a format that is slightly different (or occasionally very different) from the original.

Before we make further progress, we must look at one more definition: *news journalists*. These are a very different breed from property journalists, and they work on the main "hard news" sections of papers. Most agents or developers have little or no involvement with them at any time but just occasionally a story will crop up which means the news journalist will be in touch.

Speculation about celebrity house-buying is a good example. Although this appears exotic and rare, few central London estate agency offices have not at some time been contacted by a tabloid news journalist following up a lead suggesting that Madonna, Elton John or the latest Big Brother celebrity-for-a-day is looking to buy in the area.

Allegations of terrorist money-laundering is another example. Several property agents in major cities have faced calls from national or local news journalists on this subject, usually in the aftermath of an incident or announcement when they require rapid answers.

In chapter 3, we will discuss how to handle such no-notice media enquiries but, at this stage, the point to remember is that news journalists are completely unlike property journalists, in the following ways:

- *They are generalists, not specialists*, and will not necessarily know anything about a specific property or the property industry.
- *They work in minutes, not hours or days* because they may be writing for a morning edition of London's *Evening Standard*, an evening print run of the next day's *Sun*, or a story on BBC Five Live radio in a few moments' time. Most property writers, whether on *Estates Gazette* or the *Sunday Times*, work in deadlines measured in hours and days.
- *They will speak to you once and perhaps never again*, so if the need arises they will be ruthless, demanding and confrontational to obtain an answer. Remember, full-time property journalists will speak to you repeatedly so are unlikely to burn their bridges with you on any one story.

Against this background of how print and broadcast media operate and the functions of their staff, there are three simple questions which estate agents, developers, surveyors, architects and the like can justifiably ask:

- Why should the property industry even bother getting involved?
- What is the benefit to the industry and its individual companies?
- What do existing property players do in terms of PR?

The importance of PR for commercial and residential property professionals should be obvious — it helps them build their brand and shift their product.

Why and what to invest in media relations?

Building brands

"I've heard people say at dinner parties 'I've bought a Savills house' as if it's a type of property or a developer, which of course it isn't. But we all know why it's been said — because Savills has worked for decades at creating a successful and respected brand and, generally speaking, we all know what it stands for as an estate agency," explains Alison Dean, joint managing director of property PR consultancy Jago Dean.

She likens the process of property PR to that of selling cars. For years, the Skoda brand was generally derided for its alleged poor quality and its association with old-fashioned design and outdated production techniques behind the Iron Curtain.

"But, hey presto, that changed a few years ago because it used PR effectively to keep hammering home the message that Skoda was different now. The PR wasn't always about new models or any one feature on a vehicle. Instead, it was about building the brand and the concept that Skoda was now a good, internationally respected name," she says.

Dean is typical of many PRs in saying that commercial and residential agents are much better than developers at keeping up a continual brand prominence. It is slightly easier for agents — most have a continual throughput of properties and at any one time it is likely one or two of them produce good stories — but she says most developers are simply missing a trick by not employing PR support on a continuing basis instead of hiring them piecemeal when a particular development is being built and units need selling or leasing.

"Developers tend not to understand this. They spend literally tens of millions of pounds on advertising individual developments but they are wholly lacking in understanding of how PR can be used to complement this to build a brand," she says.

Case study — Savills

A good example of this is the company Dean spoke of, Savills, which is an agency with a corporate financial commitment to public relations that probably exceeds any of its rivals.

On the commercial side, there is just one in-house commercial property PR, dealing almost exclusively with the trade press — *Estates Gazette*, *Property Week*, business publications and sometimes the business pages of the national newspapers, too. That is supplemented by hiring Citigate Dewe Rogerson, a major international consultancy specialising in financial and corporate communications work for clients including Fortune 500 companies, for work on a project-by-project basis.

On the residential side, there are five PRs with responsibilities stretching across country properties, London homes, new-build, international homes and residential research. These are dedicated solely to handling national press PR, with occasional television and radio queries. On top of that, regional and local public relations for Savills are handled by five small local PR agencies dotted around the UK.

Finally, it has one in-house public relations officer for Savills Private Finance, the mortgage subsidiary that specialises in large loans to wealthy clients. Unusually, this post is held currently by Melanie Bien, a highly respected former financial journalist — a surprisingly rare occasion when a journalist becomes a PR.

As with most companies, Savills officially cites commercial confidentiality as the reason not to reveal the cost of its in-house PR operation, although one insider suggests it costs around £1.15m a year. On top of that, the external regional agencies and Citigate may add a further £600,000. The firm has a rigorous approach to monitoring its PR effectiveness and to convincing the rest of the organisation that it is money well spent.

It has created a database at Wigmore Street that records every single reference to Savills in each of the UK's 8000 print publications, and this database can be tapped into by key agents and executives from any office within the Savills empire. It also gives a breakdown of references to properties within each of the areas represented by the firm's 68 branch offices across the UK, and its 21 others stretched over four continents.

To top this up, an outside cutting agency is hired to present a general total of media mentions for the firm and every Monday the previous week's national press is closely analysed by the in-house PR team, especially matching the firm's mentions against its closest residential rival, Knight Frank. This is not just an exercise in counting numbers of mentions or measuring column inches, but a qualitative assessment of content too — was it favourable or not?

Twice a year, Savills' director of PR John Vaughan prepares a report on coverage for the firm's CEO, Rupert Sebag-Montefiore.

"This is the way the firm evaluates our worth as a press office and the effectiveness of Savills' public relations strategy. We're not only selling houses but we're building a brand, and editorial coverage is central to that," says Vaughan.

Uniquely within the property industry, Savills also has a subsidiary called Adventis, whose two PRs sit alongside the Savills residential PR team at their offices in Wigmore Street, central London.

Adventis handles PR for other residential companies, mainly developers such as Galliard Homes, although it has even had other estate agents as clients in the past. These have run (at least in theory) as direct rivals to Savills. The non-PR side of Adventis undertakes other property industry marketing work, from advertisement strategies to designing show homes for new developments.

Case study — Knight Frank

Knight Frank, Savills' most obvious rival, undertakes a smaller PR operation but with similarly high-profile results.

Olivia Gallimore, Knight Frank's commercial press manager, says the role of the firm's in-house PR team is to ensure corporate information, news about instructions, deal flow and research are communicated to the media is a constant process. "If you have enough material being sent out you create your own momentum — more publicity leads to more interest from the media, which — if managed properly — leads to more publicity," she says.

The commercial press manager covers the 20-plus commercial departments and offices in central London — for example, City, West End, Docklands, Hotels, Healthcare, Investments, National Offices, Research and so on. There is a regional network of 11 commercial offices throughout the UK, each using an external PR agency managed from London.

On the residential side, there are separate PR officers for country houses, London homes, and residential development (that is, new-build homes).

Within each of the country house offices across the country there is a nominated person responsible for liaising with the country house PR in Knight Frank's headquarters in London's Hanover Square; this involves suggesting interesting properties for possible publicity, handling requests for images from print journalists, and handling most of the provision of editorial and advertising for the local press.

"You don't need to work hard to get beautiful country properties in the local press. The photographs are gorgeous so the papers love them. A full PR effort is not required, so the local offices are highly successful with their nominated individuals," explains Olivia Smith, country house PR in London.

New developments are different. "A country house comes to the market and is normally sold quickly so there's rarely more than one story you can get from it. New developments may be planned, under construction and then awaiting final sales for a year or more, so there's the need to work at various opportunities for PR at different stages in the cycle," explains Emma Stanley-Evans, new homes PR in London. She therefore employs eight regional agencies to market and publicise new-build schemes that are on the market through Knight Frank.

As with Savills, the Knight Frank monitoring involves the use of a press cuttings agency and a more personalised assessment of references to the company in key (that is, normally national) publications. Every Monday, these references are assessed on a 1 to 5 basis according to their significance and favourable or unfavourable tone towards the firm.

Knight Frank is more advanced than most residential operators in using online services created and updated by a firm called Romeike, one of the communications industry's most prolific and successful information providers (see chapter 7 for details). Romeike's *pièce de résistance* is Mediadisk, a top of the range service used by Knight Frank's internal PR operation and a small number of other property players.

Mediadisk has more than 700,000 contacts and 165,000 media outlets listed in its database, including worldwide media, parliamentary and financial analyst contacts. It can be updated daily — typically, it makes 2000 changes to its contact and outlet databases every day, it claims.

When Knight Frank recently opened a new office in the Thames Gateway, it used Mediadisk to identify the local newspapers in East London and Essex, as well as trade journalists who may be interested in this sort of announcement. Mediadisk also works with Romeike's outsourced distribution service (a way of sending press releases directly from Romeike but branded and appearing as if they emerge directly from a developer or estate agent, for example), and with Clip Source, the firm's automated news cuttings monitoring system.

Depending on the scale of allied services requested from the firm, Mediadisk can cost up to £9000 annually.

Each month, Knight Frank's London press office compiles a summary of coverage in print and broadcasting, for distribution to the firm's senior management and heads of offices around the country. Occasionally, a sought-after potential client may also be presented with examples of editorial mentioning Knight Frank in a bid to help swing instructions.

Periodically, this frenzy of commercial and residential PR is pulled together to consider its impact.

"At the centre we take a more holistic view to ensure that we achieve the right balance of messages across the firm," according to Knight Frank's head of marketing and PR, Margaret Emmens. "We have half-yearly planning sessions where we review the coverage achieved and identify those areas where more or less effort is required. Part of the management process involves planning meetings with each department and we look at PR as part of this process. There is a cyclical element, which means that some parts of the firm are able to build their profile whilst other parts are quieter."

Case study — Carter Jonas

Despite having a turnover of only £16m and a fraction of the volume of instructions of the largest players, Carter Jonas can use media relations to even the playing field. The firm is as long established as the likes of Savills and has similar roots, with over 750,000 acres of agricultural assets in its portfolio.

But it has built a third of its business in the residential market over the last decade. With good quality media relations programmes, it can generate coverage of a similar quantity and quality as the country's largest agencies.

Marketing manager Ann Fairweather says: "Having worked in both commercial and residential, it is apparent that the residential property market is much more media savvy than their commercial counterparts. The residential market has grasped the huge benefits of being open with and helpful to journalists and the importance of good quality photographs and a carefully thought through story line. Residential agents know that as 'Boards breed boards' so 'Inches incite instructions'. We have received many valuation instructions as a result of media coverage on particular types of properties.

"On the commercial side of things it is a different story — you have to provide incentives to get them to focus on providing material for the press. The commercial players are much more secretive and play their cards much closer to their chests. They don't seem to recognise that third-party endorsement — especially from a journal as authorititive as *Estates Gazette* — has real value in the market."

Fairweather says brand recognition is notoriously difficult to measure but she has evolved one distinct way of judging its progress.

"One measure we use — the number of times that the firm's name is correctly entered into internet search engines — has shown that in one year accurate spelling has more than doubled from 4000 to 8450. Looking at the advertising equivalence of the media coverage obtained in a year, the value is around £1.6m and this is achieved with some time from myself, a full-time, in-house media relations executive and a cluster of five relatively small PR agencies serving the firm's 17 offices," she says.

Case study — King Sturge

Paul Gray, King Sturge's marketing partner, says the firm's 20 regional offices use a combination of in-house and external consultants in a way that allows them to adapt to what works best locally (in the next chapter, we will consider this idea of a "mix" of PR support in more detail).

He says the PR works at three levels:

1.   "Regional deals and instructions for the property and the local press promote the King Sturge brand and our local property commentators. We have established effective operations in this area over the past 10 years.

2.   Organised at a central, national level we promote our national market specialists in areas such as residential or industrial markets or even specific business issues. We started this five or six years ago and there is always much more we can do.

3.   The strategic level, where we aim beyond the property media to the national and financial media. Here we are aiming at what was once called "the surveyor in the boardroom" — a voice for property at the most senior level of business decision making. Twenty years ago there was a handful of property specialists at the leading national press and it was relatively easy to keep in touch with them and they had a deep understanding of the property sector. It is much harder now as property has been absorbed into the overall business brief, all our competitors are targeting the same few business journalists whose attention is also sought by the plethora of other consultants in management, human resources, technology and marketing arenas. However, we believe that a couple of mentions in, say, the *Financial Times* has greater value than several mentions in the specialist media when it comes to gaining attention in the Boardroom."

In the residential and commercial property sectors, the likes of Savills, Knight Frank, Carter Jonas and King Sturge represent something approaching the pinnacle of the industry, both in terms of expenditure on PR and quality of staff and subsequent coverage.

Other large national agents often struggle to keep up, although in autumn 2005 Hamptons International — an agency operating at the mid to top end of the market selling houses across London, the Midlands and southern England — dramatically increased its PR effort to five people and may feature more strongly as a result in the future.

At the other end of the scale a few estate agents — sometimes even those well-resourced because they deal with mass-volume sales or top-end properties in high-value locations — do not bother with PR at all, although these are fewer and farther between than ever before.

For the vast majority of agents, there is little doubt that PR effort and expenditure works, although commercial confidentiality means there is little hard data to prove just how well they work.

Most agents can give direct examples — highly specific, but persuasive.

In October 2005, for example, a veteran rock musician with a house to sell switched from one estate agent to another after the property allegedly languished on the market for some months without finding a buyer. The reason for the change? The "new" agent claims it was because its PR department had strong links with the national press and could get coverage for the property in a national paper just as the veteran's latest album was being released.

As we have seen, many major estate agents' in-house press offices or outsourced PR agencies monitor the entire mainstream property press, and measure their own or their clients' exposure against that of rivals, but such information is rarely made public.

The only recent survey made known to the broader property industry was in summer 2004 when a now-defunct PR firm, Property Publicity, studied the national newspapers' property sections for three months.

It concluded that just five estate agencies (all with substantial PR support either from in-house teams or the hiring in of third-party PR agencies) received almost a third of all national newspaper coverage.

Out of 1297 properties for sale featured in editorial stories in 10 national newspapers over the three-month period, 399 belonged to the "famous five" — Savills (then called FPDSavills), Jackson-Stops & Staff, Knight Frank, Lane Fox and Strutt & Parker.

"These guys can punch above their weight simply because they have huge public relations departments fighting on their behalf. It's not the journalists' fault — it's just that the PRs are feeding the editors such great stories, they have no reason to look elsewhere," says Eric Dixon, who ran Property Publicity.

A further 15 estate agencies amassed a total of 298 properties for sale, which means that, in total, just 20 agencies achieved over half of the editorial coverage available in the three-month period under scrutiny.

Exposure helps you win new business

Savills' John Vaughan says there is a virtuous circle of activity.

"There's an unspoken code, really. Of course advertising and editorial are separate and we'd get short shrift from any newspaper if we leant on it by saying 'print this story or else we'll pull our advertising'. But at the back of an editor's mind there's an awareness that property advertising is a huge earner for each newspaper, so there's a likelihood that agents will be included in a story — either a comment or a property," he says.

"Our PR effort and all advertising is paid for by our clients. The quid pro quo for them is that Savills in turn gets exposure in editorial. Journalists and PRs are the foot soldiers in all of this; we're the worker ants providing material to all of the property sections that have proliferated in recent years. That's quite a good rounded circle, but it only works for a company if there's a PR effort to undertake it."

These days, public relations even enters the pitching process when an estate agent tries to win a client's instruction to sell a property.

In June 2005, Savills had 434 properties on its books priced at between £1m and £3m, a further 34 priced between £3m and £5m, 11 between £5m and £10m, and nine at over £10m. "You can be pretty sure at each of these pitches the agent talked about public relations to help promote the property. Depending on what a property is like, the agent will produce statistics, cuttings and examples of what we've done for comparable properties in the past. Sometimes — increasingly often these days — there may even be a member of the press office there to explain it in more detail. We can't promise we'll get coverage, but we can promise we will try to get coverage, and that can be the difference between winning the instruction or losing it," claims Vaughan.

Alistair Elliott, head of national offices at Knight Frank, says the overall brand-building effect of PR is influencial to win new commercial business too.

"Clients like to think they are advised by agents who are well regarded and they get reassurance from seeing us mentioned frequently in the media. It helps us to build the Knight Frank brand. We are fortunate in that we have a good deal flow from departments and offices, which helps us to maintain a high profile. Yet when one market is quiet we have other markets that are active so we can use work in those areas to support our media profile," he says.

"Marketing and PR are part of the lifeblood of the commercial property market. With each development, you need an integrated campaign with advertising, events, sales activities and PR. They work together — adverts can generate press enquiries, the media can attend events, PR prompts interest from tenants and other agents and so on."

Advertising equivalence

Perhaps the single most persuasive argument for property companies to employ PR is demonstrated when press offices translate editorial comment into the equivalent of advertising spend.

Hamptons International told clients in 2005 that "last year our press office achieved substantial editorial coverage valued at over £1,500,000". That was a good return for an in-house press office that, at the time, consisted of only two members of staff, although many of its competitors claim even better results, some with a similarly modest press office.

Savills, with what has, until recently, been a much more sizeable press office and larger network of estate agency branches, says its advertising equivalent is £10m a year on the residential side alone — "and that's a pretty conservative estimate, it's almost certainly rather more," says John Vaughan.

Residential publicity opportunities

We can scarcely miss the importance of houses, interior design and even gardens to the national newspapers and magazines. In summary, these are the PR opportunities that exist for residential property professionals.

National newspapers — in 1997, there were only four regular property sections in national newspapers each week, and now there are 13. These sections are the well-known daily and Sunday titles, almost all of which have substantial dedicated pull-out parts carrying property editorial and advertising (the *Sunday Times* at the weekend and *The Independent* on Wednesdays are typical examples).

These sections will cover a mixture of housing market information (usually heavily researched and using clearly objective sources), lifestyle property information (covering a wide-ranging agenda from celebrities' homes to how amateur developers make money from property, through to trends such as downsizing or the growth of specialist niches like retirement housing). Many of these sections also have columns, pages or sections on interior design, furniture and gardens. Of growing interest to almost every newspaper is the rapidly increasing interest in Britons buying homes overseas.

As we will see, journalists also often want case studies (individuals whose activities demonstrate a wider point) and objective research to back up many of the stories that appear in these sections.

Local newspapers — these are the weekly or daily, morning or evening, paid-for or free newspapers common in most parts of the UK. Most have a substantial property section free with the newspaper on one or more days of the week — in some cases, the section will be wholly given over to advertising properties for sale or rent. Any with some editorial will be (in contrast to the nationals) very "soft" in the information they give — even to the point of allowing estate agents to write the editorial.

Lifestyle magazines — even *Hello!* has had an occasional property section, while longer-running titles ranging from *Country Life* and *The Field* to *Time Out* have had related sections that provide opportunities for appropriate properties to hit a focused market. Localities have lifestyle magazines too (*Hampshire Today* and *Devon Today* are sister titles and good examples of this) where property advertising and editorial occupy many pages.

Consumer magazines — there are relatively few of these dedicated solely to property but the likes of *A Place in the Sun* — a tie-in with the Channel 4 series of the same name — is one of the best read. Many of these concentrate on new-build homes in the UK or abroad, and target relatively inexperienced, young buyers as their market with plenty of "how to buy" and "how to get a mortgage" guides.

Residential property trade publications — there are very few of these because most such publications, like *Estates Gazette* and *Property Week*, are chiefly commercial. However, even these have occasional pages given over to house building and related issues.

Internet — in the residential field there are simple property sales websites (like *www.rightmove.co.uk*), there are property story sites (like *www.channel4.com/4homes*), advice sites (like *www.periodproperty.co.uk*); in commercial property there are general sales and leasing sites (like *www.propertymall.com*), niche sites (like *www.devono.co.uk* given over to London office space) and this form of communication has been strongly supported by the commercial market; and there are hybrid sites (like *www.insidetrack.co.uk*) for property investors across both residential and commercial.

Some websites carry editorial as well as advertising, although most are funded by revenue from sales and rentals. Those sites that are operated by and promoting only one company are used jointly to sell stock and promote the firm's brand (for example, *www.kingsturge.co.uk*) while portals are "umbrella" access points to the internet, as they usually pull together content based from a range of other websites (for example, *www.primelocation.com* advertises homes for sale over £500,000 from more than 100 contributing estate agents, each of which will have its own website).

The "visibility" and impact of a website or portal can be enhanced by technical features; these include using key words registered with search engines such as Google or Yahoo! to increase the likelihood that a website is named in response to a search, or electronically linking one site to another that is complementary — for example, a removal firm's site linked to an estate agent's.

Radio — occasionally a story on the housing market, a celebrity house purchase or a business report on a commercial sector will interest radio stations, usually national speech-based ones such as Five Live and Radio Four from the BBC, or some of the independent local radio "talk" stations based in large cities. A programme such as *You And Yours* on Radio Four is a good example of this. Interviews will be fixed up in advance (sometimes only with a few hours' notice, however) and may be on the telephone or in a local studio where the sound will be "patched" into a national programme. Very occasionally, radio news bulletins hold live interviews with people in the housing market but usually such an interview will be recorded and, typically, just 35 seconds of edited comment will be used.

Television — as with radio, any television interest will be triggered by the market fluctuating or a celebrity buying but, unlike radio, the interviews are a great deal more complicated and time consuming to conduct.

Again, expect only 30 seconds or so to be used; but the interview may be arranged the previous day and the camera, lights and make-up may take an hour of setting up with a "property" backdrop requested — an estate agent's office or a new development is typical. There may even be rehearsing of the interview, too. As with national papers, television loves a turgid market-type story to be made "human" by case studies of would-be buyers or sellers feeling aggrieved at being gazumped, for example.

Although television exposure for property professionals will be rare, the power of the medium is such that it deserves to be treated seriously. There are very few of what, later in this book, we call "property rock stars" who combine professional expertise with sufficient charisma and X-factor to make more than the occasional television appearance.

As with most creative jobs, there is no course (or book) that will guarantee you success in this field, but the experiences of some of television's leading property gurus may be salutary.

Case study — *Garrington Home Finders*

For 10 years, Phil Spencer has run Garrington Home Finders, an upmarket buying agency for wealthy clients across the UK. In 2000, he was invited to be a behind-the-scenes consultant on a new show helping couples locate ideal properties.

"It was just a half hour chat to explain how home finding worked. But then I seemed OK in front of the camera and everything took off," explains Spencer, who is now the leading man of British property TV, fronting five series of *Location, Location, Location* with Garrington colleague Kirstie Allsopp.

"Even so, I'm a property finder first and foremost and hope to be one long after I've stopped being on the telly," insists Spencer. He says anyone working in the medium must know their property stuff first and foremost — TV skills can come later.

"The voiceover at the start of *Location, Location, Location* is scripted but everything else is unscripted. We follow the action and respond to the couple we're dealing with, the property they want and the budget they've got. I help people spend hundreds of thousands on the biggest purchase of their life. It's my job even if the camera isn't there in the corner. You've got to know your stuff to do that," he insists.

Case study — Property Ladder

Sarah Beeny admits she "never bothered much with formal education" and flunked a drama school audition. Instead, she self-trained in the building industry and, at 24, set up a development firm with her brother and boyfriend. She had never broadcast, before beating 200 others to become front-woman of Channel 4's *Property Ladder*.

"You absolutely must know what you're talking about, otherwise you won't have the confidence. Researchers are great but you've got to think for yourself. I find things out but have no patience with long explanations. I try to convey that to the viewer but not in a way that makes it sound like a club — it's got to be accessible," says Beeny.

She says television has been "amazing, fun, remarkable", but admits it's not really her job. "I'm a developer," she insists.

Most *grandes fromage* of property television have similar stories to tell.

Kevin McCloud runs a product design practice, specialising in lighting and furniture. He has worked on interiors for Ely Cathedral, Edinburgh Castle, two European palaces, the Savoy and Dorchester hotels and dozens of private homes — in between, he presents *Grand Designs*.

Kirstie Allsopp, the polka-dotted property doyen who presents with Phil Spencer, worked for homes magazines and interior design practices before joining Garrington Home Finders. Amanda Lamb of *A Place in the Sun* first found fame as the face of Scottish Widows on television adverts and has presented several non-property shows, but before all that she was an estate agent.

The simple common thread for them all, as Spencer emphasises, is that they know their property industry first and learn their media skills later. For readers of this book the story is the same. You almost certainly know the property industry inside out — now it is time to learn about media relations.

Commercial property publicity opportunities

In the commercial property sector, there is an almost equally diverse range of opportunities to generate publicity.

Business sections of the nationals and Sunday papers — Whether you are focusing on the investment market or on the occupier market, information regarding changes or trends in these markets will be of interest. Many surveying practices and property-related consultancies will invest significant amounts in research — and this will be of interest to these media. National media will be particularly interested in research that has wide appeal to either the business sector or the majority of home owners. Also, the opinions, forecasts and advice from your key people — providing they are of interest to a sufficiently wide audience — will also find a home here.

Property trade press — The key property weeklies and the monthly magazines focused on the commercial property sector will clearly present opportunities. They will all have a list of advance features that they plan to cover through the year, which will cover a variety of locations, topics and issues. Many of these magazines also have an online version, which is a source for publicity as well. At the last count there were 20 magazines covering commercial and industrial property, a further 12 listed under real estate, two on relocation, seven on facilities management/premises administration and 23 for trade and technical readers in housing. There are also 25 trade and technical magazines on architecture, 18 on civil engineering, 10 on town and country planning and four on surveying. There are also over a hundred magazines focusing on building, construction and materials — many of which present opportunities for property angles. There are also publications on subjects as specialist as retail property and valuation — there are even property opportunities among the 60 environmental specialist magazines.

Business magazines — There are a wealth of general business magazines aimed at the chief executives, finance directors and other senior management at both large and small companies. For example, there are more than 40 general business magazines (including publications as diverse as *The Economist* and *Financial Director* at one end to *Industry Europe* and *Midlands Today* at the other), 70 on general management, 50 regional business magazines, 50 focused on industry, 40 magazines published by Chambers of Commerce and other business groups and over 20 for small businesses. Information that helps senior executives assess the future impact of rents and other property costs, ideas on how to outsource property services, increase cash flow by innovative use of leases and/or flexible premises deals will be of interest to these media.

Other industry sector trade press — Property issues, particularly for occupiers, will be of interest to a number of market-specific magazines providing they are relevant to and targeted to that sector. While you will need to have a good knowledge of these sectors and the media in order to tailor your property-related stories, there are plenty of opportunities for those that have a strategy focusing on particular niche markets. For example, there are over 100 on retail trades and merchandising, 60 magazines for the computing industry, and nearly 40 on advertising.

Commercial property portals — There are a number of websites that contain significant information about the commercial property sector and provide links to many other property-related sites. These sites usually have news pages and benefit from having high traffic rates among fairly targeted markets. Good examples include *www.propertymall.com* and *www.shopproperty.co.uk.*

PR Structures for Property Organisations

In chapter 1, we touched on how different organisations invest in media relations expertise — in-house professionals and PR agents. In this chapter, we consider in more depth the ways in which a property company can obtain and use the relevant media relations expertise.

In essence, there are three options:

- Use one of your existing marketing, business development or other in-house professionals to manage media relations as part of their role.
- Appoint an in-house specialist media relations expert (in effect, set up a press office).
- Engage the services of an external media relations consultant.

Of course, many companies employ a mixture of these options. It is not uncommon to find an in-house media relations expert managing the day-to-day work of conveying information about the organisation and its products as well as an external media relations consultant to manage high-level communications with the most senior city editors, or to manage crises or to focus on particular product campaigns.

There is also an evolution aspect. Companies unfamiliar with media relations will often retain an external media relations expert at the start. As their own experience and expertise (and comfort) with the media relations process grows, and as they spend more with the external agency, they may consider taking some or all of the process in-house. Similarly, once an in-house media relations team is working at full capacity on day-to-day matters, they may use external advisers for specific projects or objectives. The loss of a key in-house media relations expert may prompt the use of an external agency as an emergency measure. The loss of a key member of the account team at the media relations agency may drive the company to ensure that the expertise and critical relationships are managed by people employed within the company in future.

Advantages and disadvantages of in-house versus external agency

The advantages and disadvantages, and the costs, differ significantly with these options. The "right" way to resource your media relations activity will be dependent on a number of factors including: your particular objectives, the size of your property organisation, the existing in-house media relations knowledge and expertise you have, the nature of the media relations task (and its length), the media you are targeting and, possibly most important, your budget.

In-house and external agencies have their respective enthusiasts.

Jon Shilling, a chartered marketer and until recently business development manager for engineering firm Golder Associates, defends outside agencies: "Using external consultants taps into specialist expertise. While experts within the company know what they do, they fail to understand why other people are not as fascinated by their area of expertise as they are. No external PR consultant can have the depth of knowledge of an expert, but they can spot a good story that has wider appeal."

Clive Lucking, of commercial office interior design company Area Sq, takes a contrary view: "We have found that writing a piece on 'design' or 'office trends' or 'top 10 tips on how to achieve a successful corporate interior' got coverage within the property press — a piece written in-house by industry experts is always well-received by editors. The benefits about having an in-house PR department is that they live and breathe our culture, philosophy and strategy on a daily basis. It might be more difficult for someone external to grasp that."

Table 2.1 on pp32–35 summarises some of the main differences in the approaches — although these issues are explored more fully in the remainder of this chapter.

Why integrate media relations into your marketing?

The difference between marketing and PR (or media relations) is explained in chapter 1.

While you need different skills (usually found within different people) for media relations, it is important that their efforts are closely integrated within the overall marketing effort which is, in turn, closely tied in with the organisation's overall aims and strategy.

Too often, media relations is treated as a stand-alone and separate activity and isolated from both the thinking of the most senior people within the firm and the day-to-day issues and activities within the marketing operations of a company.

While this can be a problem when there is a separate in-house media relations — or press office — function within a property organisation, the problem can be exasperated when an external PR consultancy is used. A key issue here is the effective management of the PR function and good internal communications.

So you should strive to integrate your media relations into your other marketing and business development activities as much as possible so you achieve the following benefits:

Consistent messages supporting your strategic aims and messages — You need to promote consistent messages to the media. There is a danger otherwise that the media relations activity will send different messages or out-of-date information to the media than that which is being communicated by other means, such as advertising, direct mail and personal contact.

To do this requires your media relations people to talk to and meet with both senior management and other marketing teams in the business on a regular basis. You need to trust your media relations people to the extent where you can share with them the most confidential aspects of your forward strategy. On a day-to-day basis, you need to ensure that everyone receives copies of what others are doing or plan to do and allow everyone the opportunity to provide ideas and input. As you would ensure with your internal marketing, this means that senior management must devote time and energy to regularly briefing the media relations consultants on the overall strategic thinking and advising them at the earliest opportunity of any major new developments.

More effective marketing — Marketing is far more effective and more likely to generate good results if all the different activities —

Table 2.1 Advantages and disadvantages of different PR structures

PR structure	Advantages	Disadvantages
In-house marketing or business development person	*Cheap* You are already paying the salary of this person so asking them to take on a media relations role incurs no additional cost. *No learning curve* These people are already familiar with your organisation and its products (services) so they will not need additional management time at the outset to get them up and running. *Ever present* These people are in constant contact with both senior and operational people in your business. Therefore, it will be easier for them to identify potential media relations news and features while they go about their usual jobs.	*Lack of expertise* Marketing and business development professionals may not have the necessary skills or experience in media relations. This includes writing ability as well as the skill needed to identify a story or angle. You may therefore have to pay for them to attend courses or study and this may take time. *Lack of time* They will have many other calls upon their time and it will be difficult for them to devote sufficient time to media relations to make it work well. It is likely they will adopt a reactive approach to media relations — only responding to calls from journalists rather than actively promoting your business to the media. *Lack of contacts* It is unlikely that they will have come into contact with many journalists if they have not had a media relations role before — it will therefore take time for them to identify the relevant journalists and build productive relationships with them.

In-house media relations expert

Integrated communications
These are the people who are planning and managing your other communications campaigns so you can be sure that all media relations activity will be integrated with and supportive of other activities.

Trusted by senior management
Existing staff should have a good grasp of the business and the trust of the senior management and are therefore more likely to be informed about major situations that could have a media angle and be asked to contribute their input.

Limited media relations opportunities
This is an ideal solution if you are within a small organisation and/or you anticipate little demand for ongoing media relations support.

Integrated communications
Providing the person is properly integrated into the existing marketing team, their work should support and reinforce that of others in the organisation, ensuring a consistent message is communicated to the market.

Ever present
Again, their constant presence within the organisation means that they are more likely to hear about and be able to pick up news stories.

Lack of credibility
With little media relations experience, their credibility internally and with journalists may be limited.

Lack of objectivity
After a while, the individual may find it difficult to break free from the internal constraints and views in order to effectively perform their external communications role.

Limited time
It is surprising how quickly a dedicated media relations person's time gets taken up. Often, they find themselves diverted into internal communications and other activities so that their time on media relations is limited.

Table 2.1 continued

PR structure	Advantages	Disadvantages
	Focus and co-ordination Everyone within and outside the business will know that there is a central point of contact for media-related information. This can speed up and ease communications and provide a good co-ordinating role.	*Low-level contacts* While they will have good contacts and relationships with journalists at their level, they may lack the contacts and gravitas to form productive relationships with more senior journalists and editors.
	Good day-to-day media relations With a dedicated resource, it is likely that the organisation will become used to preparing information for and working with the media, so there is a constant dialogue with the media.	*Difficulty managing key projects and crises* The daily workload can make it difficult for the media relations officer to find time for major new opportunities, big campaigns or crises.
External media relations consultant	*Expertise* They will be experienced in all aspects of media relations and an excellent source of advice.	*Expensive* Whether paid on a retainer or project basis, the cost per day can be expensive. Although the use of freelances and independents is considerably cheaper than larger agencies.
	Resources Typically, you will have an account manager dealing with day-to-day media relations and a host of other experts (with other contacts) to call upon as demand requires.	*Out of touch* Often, senior management will fail to give external consultants sufficient time and information to enable them to perform their function adequately.

Conflicts
You will need to decide whether you wish to benefit from their experience of working with companies similar to your own and tolerate possible conflicts or whether you wish them to act for you exclusively and therefore have no other directly relevant experience.

Culture clash
Sometimes, external consultants are unaware of the very different culture of their client organisations and this can lead to misunderstanding, unmet expectations and troubled working relationships.

Turf wars
Occasionally, there can be difficulties between internal and external media relations advisers. This can result in a lack of integration in your overall communications campaigns.

Lack of trust
As they are external to the organisation, the media relations consultant may not be fully trusted, which means that they do not receive adequate information about major developments at the appropriate stage.

Crisis management
In difficult times, it is helpful to be able to divert enquiries to an external source while those internally concentrate on dealing with the crisis.

Excellent contacts
Good media relations consultants will have a wide variety of contacts and relationships across a broad range of media.

Build your knowledge
If you have little experience of media relations, an external consultant can help your organisation learn how to develop productive relationships quickly and provide a great source of reassurance to those in senior management.

advertising, publications, events, newsletters, direct mail, networking and selling — work together in an integrated way to ensure that consistent messages are repeatedly reinforced through each other and by media relations.

Again, this requires a property business to think about marketing in a holistic way and considering the role played by the different tools — advertising, events, selling and media relations — for every situation and campaign. This can be an issue where an organisation has a marketing team comprised of different functional experts — for example, those with expertise in events, databases, creative websites etc and you need a good strategic marketer or a range of business development professionals who can consider the contribution of media relations to a whole host of marketing and sales objectives.

Time saved for senior management — If marketing, sales and media relations functions are properly integrated, there is possibly a saving in senior management time. It means that senior management has only to provide one briefing to all those involved and to pass on information to an internal single point of contact knowing that it will be communicated as required to all those involved. If the media relations is kept separate, then senior management has to effectively brief and update lots of different groups of marketing advisers and this can lead to inconsistency of message and a lack of co-ordination.

Time saved for everyone and better flow of information — There will also be time savings (and therefore cost savings) if the media relations function has constant access to the marketing team and all the people in the organisation. The media relations experts will have a regular source of news and story ideas and activities and information can be "recycled" for media relations purposes rather than having to reinvent the wheel each time. This is particularly important in ensuring that your media relations experts attend meetings at the same time as your other marketing and business development experts with your internal teams, otherwise much time will be wasted in separate meetings going over similar ground.

Better internal communications — Using the materials produced by the media relations professional to enhance internal understanding and communications is a cost-effective method of improving internal morale and the level of knowledge about what the organisation is doing and progressing against its goals. For example, press releases can be circulated internally when released to the media, posted on intranets and on the website — where they provide a valuable information source to clients and customers.

Chapter 1 describes how Savills, Knight Frank, Carter Jonas and King Sturge organise their PR resources. Here we consider other models that exist.

Case study: large commercial developer — Rok property solutions

"We see the media as an important tool to help us meet business objectives. The fundamental principle is that the media have a job to do — to report news and write stories of interest to their readers. If we help them to do this effectively, there should be some kind of quid pro quo — accurate coverage of those items that are important to us and their readers," explains Gavin Snook, Rok's chief executive.

"We are one of the only property companies to have appointed a brand director. Media relations is a vital tool in how we communicate our brand values to various audiences — our investors, the City, clients and partners, our people and, of course, those people we would like to work for us," he says.

He says back in 1984 he had a former journalist working on PR for Rok for half a day each week. She now works full time, co-ordinating a network of regional PR agencies working for Rok, producing an in-house magazine and working alongside an external specialist financial PR agency which promotes the firm's investor relations. The PR agencies operate on a retainer basis, to produce a target number of media mentions and stories each month.

"Our aim is to be the 'The Nation's Local Builder' with offices throughout the country. We therefore need to be embedded in the local community. We want our people to become a part of their local community — to develop trust and strong relationships through the work we do in those communities. The local media is a means that helps us achieve this. We provide our views on local issues. We try to provide practical local help. The Rok Community Challenge is an initiative we conduct in conjunction with the media. We ask our people to identify an important local project — it could be a community centre, a sports hall — then we provide the expertise, tools and materials and look to the community to provide the labour force," says Snook.

Case study: medium-sized residential developer — Galliford Try

As with many residential developers, Galliford Try adopts a mixed economy of media management for different objectives.

"We employ a specialist external financial PR firm for our corporate PR. It's not our core activity to cultivate relationships with *Financial Times* journalists, so we leave it to those who have already done so. But we're much more of a regular name appearing in the property press, the construction press and so on, where we have a continuing profile," explains company secretary Richard Barraclough.

Galliford Try tops this up with in-house PR manager to co-ordinate the national PR for the house building companies within Galliford Try (Try Homes, Stamford Homes, Midas Homes and Gerald Wood Homes, each operating in a different region of the UK). In addition to this, each of the four house-building companies has its own local PR — some are in-house, some are external agency — to handle local publicity.

"We hope we understand that building relationships and establishing a brand is a long-term activity so we do review these structures, our in-house/out-of-house mix and our contracts with external agencies, but only every three years or so when we believe we can make a sensible judgment on performance and effectiveness," says Barraclough.

Case study: commercial player — Slough Estates

Slough Estates is one of Britain's commercial property industry gems. In 1920, a group of investors bought a 600-acre site in Slough known as The Dump (it had 17,000 derelict vehicles left there after World War One). Now it has become a leading source of space in one of the most lucrative commercial corridors in the world, but also a massive employer in a socially mixed community.

"While our 85-year heritage and strong foundations are important, external perceptions of what's on offer here can easily become outdated. That's why a regular flow of information from the estate to the outside world is paramount. We have a lot to say to a very diverse audience and media relations is arguably the most effective channel of communication for this purpose," says Estates marketing executive Stanley Marek.

"Through our PR advisers, F D Tamesis, we run parallel media relations programmes for local business and community audiences and national business and property audiences, targeting their respective media with tailored information and messages. Having a dedicated PR consultancy purely for the Estate means that we can match site-specific messages to the appropriate media, working within the overall framework of Slough Estates International's corporate communications strategy."

Case study: large national surveying practice — Atisreal

Atisreal in London (formerly Weatherall Green & Smith) used a mix of different media relations resources. Situated within the firm's marketing and business development team is a dedicated press officer who co-ordinates all media relations activity under the supervision of the marketing director. The company used two external media relations consultancies — one for promoting the corporate messages to the national media and another for ensuring that all the deals and transactions are communicated to the property trade press. In addition, the media relations officer has to co-ordinate with the firm's overseas owners and branches to ensure that information about the international business is consistent across all its territories.

Case study: small surveying and architects practice — Pellings

"We know we must use the media to raise our profile and obtain coverage in the press," says Neill Werner, partner and director of architecture at Pellings.

"An ideal opportunity arose with the completion of a prestigious project for Cadogan Estates — Cadogan Hall, the new home of the Royal Philharmonic Orchestra, for which we were building surveyors and project managers. We appointed an experienced media relations consultant and thought we were in a fantastic position. A week before the official launch, our consultant briefed *Architects' Journal* and *Building Design* — they were interested in running a review with photos. Then there was the successful Olympic bid and the awful London bombings and, of course, there was no interest in our project. The story lost its impact and the window of opportunity passed," he explains.

"It's a hard lesson to learn — do everything right and still there are things beyond your control that impact your ability to get coverage. It taught us a valuable lesson though: don't rely on just one or two opportunities but take professional advice and look for interesting angles in less high-profile projects — exploring the political or financial angles rather than the pure property angles. It takes a lot of effort on an ongoing basis and there are no guarantees of success — but you really have to stick with it, and acknowledge that luck can play a part."

Skills for effective media relations

The difference between marketing and PR and media relations is explained in chapter 1.

In this section, we explore the various skills and aptitudes needed by a good media relations professional. This section should help you whether you are looking for internal people with the right skills, a potential in-house media relations officer or an external media relations consultant.

There are some core competencies you would expect to find among all media relations professionals:

- *Interpersonal skills* — The first task of the media relations professional is to win the trust and respect of everyone in the organisation (the most senior board member to the most junior post room assistant) as well as journalists. So their communication skills must be highly developed. This means that they must be good at getting on with people — whether on the telephone or face-to-face. They do not need to be raging extroverts, but a general interest in other people and the ability to talk confidently and professionally to a wide variety of people in different situations and with different media will be important.

- *Writing ability* — Although a degree in English would be a good foundation, not all English graduates have the ability to write for the media and many excellent writers have never formally studied English. Naturally, you would expect a media relations person to use grammatically correct English and have good spelling. An ability to write in short sentences and to convey complex ideas in a concise and clear manner is needed. It is important to note that the writing skills for good media relations are not necessarily the same as for other forms of marketing and promotional writing — the approach in, for example, writing web copy and direct mail and advertising, is completely different to that for media relations. Writing is important in obtaining the input as well. This does not necessarily mean shorthand skills, although these are invaluable, but the ability to take detailed and accurate notes is important for when taking a brief, "interviewing" people externally and noting journalists' or editors' requirements or enquiries.

- *Questioning and interviewing skills* — In order to elicit information from people who are unfamiliar with what makes either a good story or an interesting article, a media relations professional will be skilled at asking both open and closed questions. They will need to be confident enough to ask for more detail, further explanation, clarification or examples if they are dealing with unfamiliar material. Questioning skills will also be important for the media relations person attempting to train senior management to respond appropriately to media enquiries and interviews. If a media relations professional is ghost-writing an article (for example, so that the byline of one of your managers can appear on the article), they will need to ask a long series of structured questions to elicit the required information.

- *Listening skills* —The media relations professional will spend much time listening to others in order to identify ideas for publicity as well as absorbing information about detailed issues that may be beyond their knowledge and expertise. Active listening requires the ability to maintain an open mind, avoid physical and mental distractions, take detailed notes, seek regular clarification and provide continuation prompts for those they are listening to — either on a one-to-one basis or in group environments.

- *Empathy* — The ability to see things from another person's point of view is essential for media relations professionals. They must have a deep empathy with journalists — and the different tasks they undertake whether this is on the news desk or in preparing interesting and topical features. They must have the ability to think like those who are reading the media to which they are contributing material and they must be able to understand the constraints, concerns and different ways of thinking that might be adopted by those who they are trying to promote to the media.

- *Patience* — A good media relations professional will be patient and understand sometimes that they must invest a lot in both internal and external relationships before they bear fruit. In the property world, where media relations is often misunderstood and treated with suspicion, they will need patience to coax people to share and prepare material for the media and manage their disappointment if things do not work out as quite as expected.

- *Attention to detail* — Whether recording facts and figures, complex legislative information, the views and opinions of senior people within the organisation or standard material about your organisation and its products and services, the media relations professional must get the detail right. Typing and grammatical errors — often generated as a result of being under too much pressure or trying to achieve too much within tight timescales — can seriously undermine the credibility of a media relations professional both within and outside an organisation.

- *Organisational ability* — Managing media relations activity requires strong management and organisational skills. Standard material on a wide range of activities that has been approved must be kept up to date and disseminated throughout the organisation. Detailed lists of topics and potential media and information about the preferences and past dialogues with journalists must be kept easily accessible. Minutes of numerous meetings about forthcoming activities and opportunities need to be managed to ensure that the high-priority activities receive the attention they need at the appropriate time. A media relations person has to maintain so much information to do their job efficiently that they must be masters of organisation and at their best when juggling multiple projects and deadlines.

Qualifications

While there are a variety of qualifications that can help you assess the quality of the media relations professional you are considering, they are no indicator of their likely effectiveness. Experience and track record are the only real indicator of media relations effectiveness.

- *University degree* — It is possible to do a degree in journalism (and, more recently, a degree in public relations). Usually, these courses are highly practitioner-based to ensure that the appropriate skills have been developed.
- *Chartered Institute of Public Relations (CIPR)* — This is the professional body for individual public relations practitioners. However, it covers the full range of PR activities and not just media relations. There is a professional body for the actual PR consultancies — the Public Relations Consultancy Association (PRCA).
- *Chartered Institute of Marketing (CIM)* — While this is the gold standard of marketing qualifications, it considers marketing communications and public relations from a strategic point of view. There is no specific course content aimed at helping people develop the skills necessary to be an effective media relations practitioner.
- *Communications Advertising and Marketing Foundation (CAM)* — These qualifications are more oriented to marketing communications and PR than CIM.

You will also gain some insight into the likely skills of a media relations professional by considering the nature of the professional bodies to which they are affiliated. The most common ones are:

- *National Union of Journalists (NUJ)* — A trade union for journalists that sets out the fees and terms by which members are employed or commissioned; however, fewer than 60% of working journalists are in this organisation. A far smaller proportion of freelance journalists — who as we will see in chapter 3 are the most active within property writing — are members of the NUJ.
- *Chartered Institute of Journalists (CIJ)* — A less militant professional organisation that provides particularly good support for freelance

journalists; however, it is believed that fewer than 2% of working journalists are in the CIJ.

Knowledge and experience for effective media relations

The experience and knowledge of a good media relations professional will be varied and highly dependent on their background, their past roles and responsibilities. Although you are unlikely to find all of the possible experiences and knowledge in one individual, it might be useful for you to consider which are most important to your organisation and the specific media relations challenges your organisation faces.

- *Commercial (B2B) or consumer (B2C)* — This distinction is vitally important for the property industry. For corporate PR support, you will need a PR that is experienced in B2B media relations — dealing primarily with the trade and technical press. For residential developers and those targeting consumers, you will need a PR with experience in the consumer press.
- *Journalism or PR roles* — Some argue that the best experience for a media relations professional is to have trained and worked as a journalist. Not only will they have a real insight into how the media operates and its requirements, but also they are likely to have a considerable number of friends and established contacts in the media. The downside is that sometimes journalists find it hard to adapt to the corporate environment and may miss the variety of different writing tasks too. Someone who has always worked in PR is also likely to have a really good understanding of the media.
- *Property sector expertise* — This is a contentious point. There are some who would argue that a good media relations person in the property sector really needs extensive and in-depth property knowledge. However, there are many excellent PRs who can absorb the required property information quickly and get to grips with the property sector press fast. For residential property companies, good experience in the relevant consumer markets may be more valuable than property experience. Non-property PRs have the added advantage of experience and contacts with a wide range of media outside the property trade media. However, in complex areas, such as property investment or highly technical

property services, it is probably better for them to have experience of the property sector.

- *In-house or agency* — A PR who has spent the majority of their time in a PR agency will be used to juggling many different projects and priorities and be quick at absorbing new information and new ideas. However, they may find it hard to adjust to an in-house role that lacks the variety of a PR agency role and the relative loneliness of being a PR among so many property professionals. On the other hand, someone who has worked at an agency is likely to have had experience with a wide variety of clients, sector and media.

To summarise, while media relations professionals come from all types of backgrounds, there are, typically, three environments in the property sector that they are likely to have come from:

1 *Journalists* — These are people who have a journalism degree and spent the early part of their career working as journalists in a variety of media. Clearly, these people will have many of the above competencies. But sometimes they will find it difficult to adjust to being in a corporate or PR agency environment and comply with the various policies, procedures and systems that are needed for effective control of media relations.

2 *In-house press officer* — These people have studied and trained in public relations. They may have a degree in English or some other subject and then completed additional studies in public relations. Sometimes, they have moved from university into a PR agency and learned the trade by working on a variety of client accounts. Recently, we have seen the emergence of a degree in public relations but usually they have completed post-graduate training through the Chartered Institute of Public Relations or one of the other media relations bodies. Sometime they may have been in a broader marketing communications role and decided to specialise in media relations.

3 *PR agency* — Some media relations professionals move from PR agency environment into in-house press officer roles. When they have been on both sides of the fence they are likely to have a greater appreciation of the needs and challenges of each and thus be better able to balance the use of internal and external resources.

Appointing an in-house PR officer

The starting point must be an overall strategic marketing plan and a clear marketing communications strategy that shows how media relations will support the achievement of the company's overall objectives and specific targets that must be achieved.

The balance between corporate PR and product PR (see chapter 1) should be addressed at this stage. You should also give consideration to the level of seniority of the person being appointed — you will need someone with far more experience (and grey hairs) if they are to operate mostly at board level, whereas if you are more concerned with maintaining a regular flow of information between the organisation and the target media then a more junior person could be employed.

Then you need to develop a job specification and also a person specification. An example job specification is shown in table 2.2. Make sure that you modify it to reflect the specific needs and challenges of your organisation.

Table 2.2 Illustrative example: In-house media relations officer job description

Title	Press officer
Location	London-based but regularly visiting three regional offices and numerous sites
Reporting to	Marketing director
Salary and benefits	* See below
Objectives	The primary purpose of the role is to raise the profile of the company overall and to develop and maintain strong relationships and positive coverage with the relevant national, property and other trade and technical media. The secondary, but equally important, role is to provide media relations support to the agents and in-house team promoting new residential developments.
Targets	These will be reviewed after six months but are likely to include:

Corporate PR
- Establish strong relationships with the 12 target media already identified.
- Ensure that at least one press release is issued each week.
- Ensure that we contribute at least one feature article each month.
- Arrange four media interviews each month.

- Achieve a 40% increase in media mentions in the relevant media within six months.

Residential campaigns

- Arrange for at least six media visits to the new development in the agreed launch phase.
- Arrange for at least two feature pieces in the relevant consumer magazines at the pre-launch stage.
- Arrange for at least six feature pieces in the target local/ regional media.

Roles and responsibilities

Strategic communications	In conjunction with senior management and the marketing director, develop a corporate media relations plan highlighting the key messages to convey and the target media and setting out a programme for the year and the means by which media relations will be measured.
Internal liaison	Attend board meetings and departmental meetings on a regular basis to capture ideas for media coverage and to encourage all members of the firm to generate ideas and material. Ensure all information released to the media is circulated as appropriate internally and made available in the various media, such as the intranet and website.
News releases	Ensure all newsworthy activities within the firm are adequately reported to the relevant media on a regular basis.
Features	Regularly review the forward features of our target media and ensure that the company submits suitable material in a timely manner and ensures that our expert views and opinions are featured where appropriate.
Enquiries	Establish effective media enquiry systems to ensure they are directed appropriately in-house and that a rapid response is provided to journalists.
Training and coaching	Identify the key needs for education and training in media relations skills and organise training and coaching as appropriate.
Interviews	Help prepare senior management for media interviews, attend if required and undertake the required follow-up calls.
Systems	Identify the appropriate resources (directories, online services etc) to ensure we have up-to-date media distribution lists and the required level of knowledge regarding our target media and the journalists with whom we work most often. Set up and maintain systems to monitor the media relations outputs and the results. Maintain systems to ensure early warning of forthcoming key media events.

Policies	Help with the development of the various policies in relation to media relations — who may speak to the media, the training requirements, how information is approved and checked before release and how serious situations are managed.
Procedures	Ensure that there are complete and up-to-date procedures for the release of information to the media, for handling media enquiries and for monitoring the results. Ensure that standard materials are maintained and kept up to date.
Letting and sales team support	Work closely with the various letting and sales teams on key residential campaigns ensuring that the local, property, consumer and national media are kept informed.
Directory entries	Ensure all directory entries are maintained and up to date.
Awards etc	Ensure that journalists are invited to and/or made aware of any awards or special events (such as exhibitions) that the company participates in.
Other	Other duties as required.

You should also make available a "person specification", using the guidance above to indicate the sort of person and the required experience they should have.

Appointing an external PR agency

There are a number of stages in appointing an external media relations consultant and the key to success is advance clear thinking and planning.

Link to the business plan and strategy — The starting point, as with the appointment of an in-house press officer, is to consider the overall business plan, the strategic marketing plan and the communications plans for the company overall and for specific developments or other projects.

Prepare a brief — Then you should prepare a clear brief (see the example below).

Illustrative example — external strategic PR brief

- *Introduction*
  Explain the purpose of the brief. Explain how the process for the selection of a suitable PR agency will proceed. Explain what the invited PR agencies should do in order to register their interest and pitch for the work.
- *Our organisation*
  Provide an overview and introduction to your organisation. Include a brief

history and facts and figures relating to its current operations. Indicate the key mission, aims, values and operations. Mention any critical issues that have a bearing on its past or present external profile. Mention its present objectives and strategic goals. Refer to sources of additional information, such as enclosed copies of the annual report or the website.

- *Our marketing and PR organisation*
  Describe the present marketing and PR organisation (if any). Indicate the strengths and weaknesses with regards to its impact on media relations. Highlight the key corporate or product objectives that will need to be addressed by the PR agency. Indicate the key players in the organisation with whom the PR agency will need to work. Describe how the PR agency will be supervised on a day-to-day basis.

- *Overall media relations objectives*
  Explain — in as much detail as possible — what you expect the media relations agency to deliver in terms of results and the milestones by which progress will be measured. Make it clear what specific outcomes are expected and on what basis.

- *Requirements*
  Indicate whether there any specific requirements that must be met by the agency — this may include, for example, particular experience or knowledge within key sections of the property industry, specific services (eg, media training), particular skills among the allocated account team, nominated contacts within specific broadcast, print or electronic media, media and analysis measurement systems, location of offices etc.

- *Priorities*
  Indicate those activities and results on which the PR agency is to focus in the short term. This may include aspects of the familiarisation programme, specific projects that require urgent attention, the first campaigns that need attention, a review of the perceptions of selected media, the creation of internal and external information gathering or distribution systems. It might also indicate the short-term results sought.
  Once the initial priorities have been addressed, there may be additional activities that need to be undertaken.

- *Contractual and budgetary issues*
  An indication of the intended budget should be shown here — this should cover both the fees and the expenditure on aspects such as travel, monitoring services, entertaining etc.
  Most organisations will also have some contractual issues — this will typically include confidentiality and conflict arrangements but other issues, such as reporting and assessment periods, may also be included.

- *Tender process*
  Explain how the tendering process will proceed, describe those who will be involved in the decision-making process. Outline the time frame for the

selection procedure and note any critical dates that have already been established for meetings. Explain what arrangements have been made for tendering firms to meet with members of the organisation in order to obtain further information to help them prepare their proposals.

- *Tender submission*
  Describe what information the PR agents are required to produce in order to be considered for the work. Typical proposal documents will include:
  - Views on their initial assessment of the stated requirements and the brief.
  - Proposals for how the agency will meet these requirements.
  - Estimate of how much time will be required (and thus an estimate of the likely fees and expenses) or an overall project cost.
  - The name and biographies of the key individuals at the consultancy who will be working on the account.
  - How the relationship will be established and operated on an ongoing basis.
  - How work and progress will be communicated — and results and effectiveness will be measured.
  - Potential start dates and availability.
  - Credentials — relevant experience and track record among property companies and other relevant markets or organisations.
  - Background — brief information about the nature of the consultancy, its advisers, the nature of its work and its key clients.
  - Other — whether there are any potential conflicts with other clients, standard terms of business, what other information is required from the firm, familiarisation process etc.

- *Background information*
  This might include campaign outlines, copies or analysis of past coverage, more detailed specifications of what is required, research reports showing present journalist views or competitor coverage etc.

Identify suitable agencies or freelances — To determine the consultancies that are most likely to be of interest to you, you should ask around. Seek the views of the journalists on your target media, ask the opinions of your colleagues in similar companies or at your professional adviser firms. You can look up suitable agencies at the PRCA website — but not all small agencies are listed here. You might consider a freelance PR officer, in which case the CIPR, NUJ and CIJ all maintain registers. You might also seek the views of your marketing professionals or those of other property companies. The PROFILE property networking group is where marketing and PR professionals within the commercial property industry get together — an excellent source.

Organise a competitive pitch — Most companies will wish to meet with and assess a number of PR consultancies so it is likely there will be some form of competitive tender (known as a "pitch" or a "beauty parade"). It is important to be clear about the time that is likely to be involved in such a process and the requirement for senior management to be involved.

Agree the critical success factors for the agency — You will need to identify in advance what the most critical factors are in your decision process — is it cost, proximity, experience in your area or subject, existing contacts, reputation, creativity, media specialists or other services included? Agree in advance what the most important factors are that you will be comparing the agencies against.

Meet the people who will do the work — Most agencies will have similar expertise, experience and services but a key success factor will be how well they get on with your senior management and other staff. For this reason, it is imperative that when you meet with prospective PR agencies you become familiar with those who will be actually working on your account.

Assess the agencies — As communications experts, most agencies will be able to put together an impressive pitch document. So you must consider a range of other factors when trying to compare them.

Table 2.3 Assessing PR agencies

Factor	Points to explore
Understanding	How well do they understand your organisation and its aims?
	Have they interpreted the brief accurately?
	Have they successfully identified your core aims and critical success factors?
	Have they made valid comments about the nature of your brief — any areas you may have omitted?
	Have they challenged you on your expectations or sought clarification for what you really require?
Track record	What do they consider their greatest achievements and campaigns?
	What have they actually done recently that is comparable to the work you require?
Experience/conflicts	How much directly relevant experience do they have of companies such as yours?
	What comparable projects have they undertaken before and what were their key achievements with them?
	Are they acting at present for any of your competitors?
	How will they deal with conflicts should they arise in the future?
Contacts	How familiar are they with your target media?
	How well do they know your target journalists?
	What other contacts do they have that might be useful to you?

Culture and style	How similar are they in terms of the way they work to those in your organisation?
	How well do their people get on with your people?
Enthusiasm	How keen are they to work with your company?
	How much preparation and research did they do for the pitch?
Capacity	How will they accommodate your work if they are successful?
	Have they identified the relevant staff who will be focusing on your account?
Creativity	Have they produced some good ideas for you to consider?
	Have they presented ideas and thoughts that are substantially different from the others?
Working relationship	How do they foresee the relationship developing and operating on a day-to-day basis?
	What plans do they have for learning about the company and the projects and getting up to speed quickly?
	What do they require from you, your management team and your marketing team in order to get started?
	How familiar with and prepared do they appear in getting to work on your behalf?
	How frequently do they intend to meet with you?
	How much time will they spend at your offices, at your sites and with your teams?
	How will they tackle difficult situations or problems in the relationship?
Reporting	How will they report progress and results to you?
	How frequently?
	How do they ensure that your goals are achieved?
Questions	What sorts of questions did they ask you?
	Did these questions convey a deep understanding of your business and your media relations needs?
	Do the questions suggest that they have already worked out what they need to do for you to get the media relations results you require?
Costs	What is the main basis of their cost proposals?
	How certain are they of the likely level of expenses?

To summarise, the process will typically comprise the stages outlined in table 2.4.

Table 2.4 Time scale to select a PR agency

Time frame	*Activity*
Week 1	Develop a clear brief and obtain approval for the brief and likely budget.
	Agree who will be involved internally in the selection process and ensure they have the required time and authority.
	Identify the critical success factors against which you will be measuring the various pitching firms.
Week 2	Investigate the likely PR consultancies that you will invite to pitch. Make preliminary contact to ensure that a) they are interested and b) they are not conflicted out.
Week 3	Issue the written PR brief with an associated information pack about your company and the media relations projects you wish to tackle.
	If time permits, allow informal briefing meetings with the agencies you have invited to tender.
Week 4/5	Receive the written submissions from the invited agencies. Undertake a preliminary review. Circulate to others in the decision-making panel. Clarify any points that are raised in the submissions.
	Decide whether you wish to prepare a short list or whether you will arrange formal meetings with all those who submitted a tender.
Week 6	Agree how you will manage the pitch meetings and the key questions you need to ask all agencies and specific agencies.
	Undertake the meetings with the agencies
	Debrief after each and all of the meetings. Agree who will be appointed.
Week 7	Advise all the pitching agencies of the outcome and provide feedback to the unsuccessful tenderers.
	Negotiate the fees and terms of the successful agency.
	Make the necessary internal announcements and start the induction process.

Costs

The primary cost of an in-house media relations professional will be salary and the associated benefits and overhead costs. Table 2.5 shows the range of salaries for the various levels of seniority within London and the regions.

Table 2.5 Indicative salaries of in-house media relations staff

Job title	Job definition	Lowest	Average	Highest
Head of communications (10–12 years)	Manages all aspects of the marketing communications strategy and plans. Entire team reports.	£60,000	£75,000	£82,000
Marketing communications manager (6–10 years)	Responsible for strategy and planning. May manage a small team looking after content and editing of internal and external; publications, newsletters and articles.	£55,000	£60,000	£64,000
PR manager (4–6 years)	Responsible for raising the profile of a firm or individual. Primary contact for trade and national press and experienced in advising internal clients in all areas of PR.	£40,000	£45,000	£53,000
Marketing communications executive (3–5 years)	Implementation of internal and external communications initiatives, including publications, advertising, website design and writing press releases.	£27,000	£30,000	£35,000
PR executive/officer (2–3 years)	Drafts press releases or manages an agency. Responsible for coaching senior members of the organisation and in smaller organisations the primary contact for trade and national press.	£25,000	£28,000	£33,000

Note: Data supplied by Marketing Resources (contact: Giles Taylor, Marketing Resources, 75 Grays Inn Road, London WC1X 8US Tel: 020 7242 6321 e-mail: gtaylor@marketingresources.co.uk).

Table 2.6 Example job advertisements in the property/construction sector — October 2005

Position	Description	Salary
Head of media relations	Property-related professional body	c£60,000
PR consultant	Property and finance experience. 4 years' experience	£25,000–35,000
Account manager	London. Property PR company. 3 years' post-graduate experience	£28,000–33,000
Senior PR manager	Agency specialising in property	£30,000
Account manager	With property experience	£30,000
Account manager	Property and construction PR agency. Out of town. Business-to-business media	£28,000
PR executive	Commercial property	£25,000
PR account executive	Manchester. Property PR agency	£19,000–22,000
Junior account executive	Commercial property	£18,000–25,000

In addition, you should ensure that the marketing budget has capacity for a number of media relations-related costs. The following examples are for both corporate PR and development/project PR.

Table 2.7 In-house media relations costs

Corporate PR

Press release stationery, folders etc	£150–500
Media directories (updated twice a year) or online targeting service	£300–1500
Press area on main website	From £500
Press cuttings service (large organisations only)	£300–3000 pa
Photography	From £350–800 per day
Coffees, lunches etc with journalists	From £20 per week
Press receptions and parties	From £1000 to significant
Research into journalist perceptions	From £4000
Media relations training for senior management	From £750 per day

Product PR

As above and also

Local media events	From £1000
Support for sponsorship and competition deals	From £3000

External PR consultancy costs

Senior PR consultant	From £800–1200 per day
Junior PR consultants	From £250 per day
Typical monthly retainer	From £1000
Monthly expenses (travel, post, telephone etc)	10% of monthly fee

Supervising and managing your PR resources

While you may think that once you have appointed your internal or external media relations consultant you can sit back and relax, in reality this is where the hard work really begins.

Overall responsibility — You must be clear at the outset who has the primary responsibility for managing the media relations resource — regardless of whether it is an internal person or an external agency. Ideally, this should be someone who is sufficiently senior to be able to make fast decisions and coax people to co-operate with the media relations professionals but also someone with an appreciation of what the media relations consultant will need and do. It may be a director with responsibility for marketing and promotions or it may be your marketing or business development director or manager.

Internal communications — You must ensure that everyone within the organisation knows who is responsible for media relations and their contact details. Explain what they will be doing and why and provide a channel for any concerns or questions about the appointment. Ensure that you regularly report on the media relations successes and what is required of those in the organisation who need to work with the media relations professionals.

Familiarisation 1: background research — Typically, there will be a period of time where the individuals develop their knowledge of your organisation and its people, services, markets and customers/clients. This is likely to use a significant amount of management time. Naturally, your website will hopefully provide plenty of useful information as well as your company's corporate and product literature. If possible, you should prepare a detailed briefing pack in advance containing as many business plans, marketing plans, project descriptions, CVs, minutes of key meetings etc to help them get up to speed as quickly as possible. The more information they have to read in advance, the less time they will spend in meetings getting up to speed with the basics. Another good approach is to assign a junior member of the company — or a secretary — to handle their initial questions and requests for additional information.

Administrative co-ordination — If you are using an external PR consultancy, it is important that you allocate them some dedicated internal secretarial resource to help them arrange meetings, gain access to the relevant people and obtain basic administrative information and support. While this appears as an overhead, it will be useful as the allocated secretary will remain aware of all the activities of the external PR agency and will a) save them time (which increases the amount of time the PR consultants will devote to real media relations) and b) start to build your internal knowledge and expertise about media relations.

Clear policies and procedures — Ensure that you are clear about what you expect in terms of approvals and sign-offs. There may be professional rules (for example, at the RICS) that indicate client or customer material may not be disclosed to the media without the prior written permission of the client/customer. You may wish to assign different individuals in your organisation to take responsibility for media relations in different departments or offices. There should be internal communications procedures to ensure that any major media relations activities are alerted in advance to board members. Similarly, you may need to issue guidance to all your staff internally about how to receive and direct enquiries from the media.

Familiarisation 2: meeting your people — Once they understand your business and services, it is advisable to arrange for the internal or external PR consultant to meet with a variety of people within your organisation. This will both add to the knowledge they have of your business and allow them to start explaining to people how media relations works and what they will need from them. This first round of exploratory talks will also enable the media relations professionals to assess who is particularly media-friendly and where the main source of stories and information might be. It is critical that the media relations consultants promote a professional and confident impression at these interviews and thus win the trust and co-operation of your team.

Regular meetings — Ensure there are regular meetings with your internal or external media relations person (and, ideally, with your marketing people). You should arrange to have an update report in advance of these meetings so that they can be kept brief and to the point. A standard agenda may assist. For example:

• Progress since the last meeting.
   – achievements

- – outstanding actions
- – problems/issues.
- Progress against the overall plan.
  - – successes
  - – outstanding
  - – changes to the plan/expectations required
  - – review against overall objectives.
- Key issues/campaigns to address at this meeting.
- Aims and major projects for the next month.
- Aims and major projects for the medium term (3–6 months).
- Brainstorming new ideas.
- Any other issues.

Trouble shooting — Inevitably, there will be problems with your in-house or external media relations advisers. The key to avoiding or overcoming these is open, regular and frank discussions about aims and expectations and where you are disappointed (or pleased) with the methods or results. When you stop communicating with your media relations professionals, the problems really begin.

Common problems and how to resolve them

There are some companies that enjoy a productive relationship with their in-house and external media relations advisers and never experience problems. Then again, the vast majority of companies have some teething problems. Here is an overview of the most common problems and some suggestions for how to resolve them.

Table 2.8 Common problems with PR agencies

Problem	Suggestions for resolving the problems
They are not delivering the expected results	• Confirm that you have outlined your requirements and communicated them effectively. • Review (and amend) the brief. • Ensure that your people are devoting sufficient time to the PR advisers. • Check the time scales — will it take longer for the results to appear than you initially expected?

	• Arrange a meeting where you openly discuss the issues and seek resolutions — inevitably there will be improvements that both sides will need to make.
The costs are too high	• Review the brief and their proposals.
	• Review what you have asked them to do — is it in line with your initial brief or has the role expanded?
	• Are they undertaking work that should be completed by others, such as your in-house staff or your senior managers?
	• Arrange a meeting where their time input is reviewed.
	• Consider focusing their roles on a more limited range of projects.
	• Arrange a meeting and outline the issues so that a resolution can be identified.
Management and other staff do not provide enough time or information to the media relations professional	• Ask the media relations consultant to specify what is required and from whom.
	• Ask senior management to speak to the individuals involved and explore the issues.
	• Assess whether other individuals in the organisation can provide some of the time and information required.
Poor-quality coverage results	• Explain to the media relations professional why you are disappointed and explain more precisely what you hoped would be achieved.
	• Provide specific examples of what you wish to achieve — refer to examples by competitors if this helps.
	• Listen to their views on why poor-quality coverage is occurring — consider what changes you can make internally to overcome this.
Initial results were impressive but not much is happening now	• Look at what happened initially and what is happening now — consider how the initial impetus can be re-ignited.
	• Undertake a detailed analysis of what happened and what resulted. Clarify the controllable and uncontrollable factors.
	• Arrange a meeting to explore the reasons — sometimes the media relations professional has been diverted onto other activities, sometimes they are working on a longer-term campaign, other times their internal source of news and ideas has dried up.

An inaccurate or negative piece of coverage occurs	• Investigate what happened and why — review your procedures.
	• Look at the situation in context — is it one minor piece in an overwhelmingly positive set of results?
	• Take time to assess the true impact of the piece — often people overreact in the immediate aftermath of a piece.
	• Develop systems and procedures to prevent it from happening again.
Omitted from a critical feature	• Confirm that you had identified the medium and feature as important.
	• Ask about the forward features monitoring process and the level of awareness of the particular type of feature.

What Journalists Want and How You Can Provide It

A journalist is someone who writes and researches stories for any publication ranging from a free local weekly paper to the revered *Sunday Times*. If the journalist writes for magazines, these could range from an in-house company magazine to the hallowed pages of *Newsweek*. Essentially, the process is the same even if the stories are completely different — a journalist spots a story appropriate to the publication, convinces the editor it is worthwhile, then researches and writes it.

So far, so obvious, but note the "convinces the editor" part of the process.

Most publications are a pyramid with the overall editor at the top. As we have seen, in the case of national newspapers, beneath him or her are section editors (ranging from news through business, sport, features, property, motoring and so on). As the pyramid descends and broadens, just below section editors come staff journalists if there are any; then come freelances who are not directly on the payroll of the newspapers but may be used regularly or occasionally, as required.

This chapter deals almost entirely with journalists (staff and freelance) rather than editors (of entire newspapers or just of the property sections) because, in the main, it will be journalists who contact property professionals in the commercial and residential disciplines, and it is the journalists who will need convincing by PRs that a story is worth writing about. The only way to do that is to know the editorial market.

The editorial market

Judging the editorial market

A good way of calculating what journalists look for is to analyse the publications or outlets for which they write. There are literally thousands of publications that include something related to property. Do not discount television and radio programmes either — there were seven stories about property on Radio 4's *You And Yours* programme in August 2005 alone, most involving company spokesmen putting their position on the likes of the residential market, property investment or key worker housing.

By way of an analysis, let us look in detail at what has become the big prize for residential estate agents and developers — getting editorial in national newspaper property supplements.

The reason why this is sought-after is simple: editorial coverage is free but carries more influence than comparable space carrying paid-for advertising. A journalist's assessment of a new development, or a house type, or the property market is considered to be neutral and authoritative. On the other hand, estate agents or other industry insiders commenting on the same things are generally considered less impartial.

This is natural. If agents or insiders came out in favour of a specific development or optimistic about the housing market, the public would quickly resort to the old phrase "they would say that, wouldn't they?".

Estate agents and developers seek to quantify the value of the editorial coverage they achieve against the cost they incur achieving it. This is a simple calculation of how much the firm would have paid to get advertising space as large as its editorial coverage (say, £2m) compared with the cost of its public relations and press office effort (say, £1m). This is known as advertising equivalence.

But column inches and the advertising rate card equivalent is not the sole measure, because the way property as a subject has been covered has altered dramatically in recent years.

For example, those operating in the commercial property world have found their subject is now an integral part of the business and finance pages as well as the specialist trade press, and so the journalists they deal with are harder-nosed and higher-calibre than they were in the 1980s and 1990s.

The same has happened, but with even more vehemence, in residential property.

In the past, you were given a picture of a 'jolly nice property', taken to lunch, and then the journalist wrote about it. Those days are long gone and property is seen as a respected journalistic discipline like finance or sport. The problem is the property industry is about 40 years behind the times when it comes to PR. Many PRs, house builders and agents have not cottoned on to this new land where we all live.

This is how Cheryl Markosky, one of Britain's most prolific property journalists, sees it.

Markosky refers to the days before the late-1990s when very few newspapers had property sections and those that did were poorly resourced and not taken seriously by other journalists or even the public.

It was not uncommon for lead stories to be nothing more than rewritten estate agents' details about a particularly fine house that was for sale. The agents' photographs were used too, and more often than not the writer of the story never set foot in the property before putting pen to paper.

Now all that has changed. Newspapers rely on property sections for much of their advertising revenue, property sections attract serious journalists to contribute (two former BBC reporters write each week for the *Sunday Times'* Home section, for example), and the public is more interested in property than ever before.

As a result, the property editors want a variety of information. Hard market comment is required, for sure, but human interest angles are also needed to interest readers.

Markosky continues:

The few property PRs who are good, understand what we want and educate their clients accordingly. The not so good PRs do whatever their clients demand in order to pull in the money. The irony is that these get relatively little or no national coverage because they are trying to play tunes from the client's song sheet. Surely it is better to come up with the stories we want — and then everyone is happy.

We will return to how, in terms of PR, "what the property industry client wants" differs from "what the journalists want" later in this chapter.

Case study: Analysis of one week's national newspaper property coverage

In the meantime, it is a useful exercise to look at one week's national newspaper property supplements to see how they differ in terms of the types of story they cover and, perhaps more crucially, how alike they are in the way they cover them.

We looked at the papers in the week beginning Wednesday 13 July 2005 to capture *The Independent*'s mid-week property supplement, and running up to the glut of supplements on the Friday, Saturday and Sunday.

Each story in each property section was categorised as being one of three types:

- *Market story* — that is, a story mainly or wholly about the state of the domestic or international residential market and associated aspects like buy-to-let and investment purchases. These are usually backed up by independent, hard facts and figures.
- *Pure property story* — that is, a story mainly or wholly about a home on sale or a development just announced or being finished, or sometimes extended to cover a geographical area and looking at the properties on sale within it. On occasions, this sort of story can be about property interiors and design as well as the shells.
- *Lifestyle story* — that is, a story that is not so much about properties but more about how people live in them or use them. Sometimes the people are celebrities but, more commonly, they are ordinary sellers or buyers with an unusual or epic story to tell.

The analysis produced these results:

Table 3.1 Analysis of one week's national newspaper property coverage, July 2005

Property supplement	Lifestyle stories	Market stories	Pure property stories
Sunday Times	7	4	12
Mail on Sunday[1]	9	2	5
Sunday Telegraph	3	1	3
Sunday Express	1	0	4
The Observer	1	0	3
The Business	0	0	1
Daily Telegraph	2	1	5
Financial Times	0	3	2
The Guardian	2	0	2
The Times	4	2	4
Daily Mail	2	2	0
The Independent	4	4	5

1. *The Mail on Sunday* property section is only distributed in the Home Counties and not nationwide.

So far so good, and no surprises. Anyone who thought the *Sunday Express* was going to give incisive analysis of the property market does not know its readership nor what sells newspapers; likewise, do not expect to open the *Financial Times* and see stories of how two pensioners have retired to the Costa del Sol.

But what is less predictable and more interesting is to look at the treatment of these stories. Understanding this begins to get to the heart of helping industry insiders and their media advisers to judge what sort of story will get picked up, by what journalist and for what publication.

One simple insight is gained by looking at the cover page of each property supplement.

Editors of sections know this is the make-or-break page — will people even bother to open the section in the first place if the cover appears dull, or echoes a story they may have already heard on television or radio? Therefore, the cover page has to appeal widely to its target audience, but be distinctive. In our analysis of that week's supplements in July 2005, we looked at the main photograph and the nature of the lead story for each title.

Table 3.2 Analysis of lead pictures and stories in national newspaper property sections, July 2005

Property supplement	Cover: main picture	Cover: main story
Sunday Times	Young couple	New owner of large estate
Mail on Sunday	The Hamiltons	Tour of eccentric homes
Sunday Telegraph	Woman in the sun	Buying French homes
Sunday Express	Woman	Ex-nun selling former chapel
The Observer	A family	Disputes over extensions
The Business	Person with laptop	Buying homes on the internet
Daily Telegraph	Diagram of home	How to make homes eco-friendly
Financial Times	Person driving car	Buying homes across Europe
The Guardian	Man in hot tub	Does a hot tub enhance a home?
The Times	Two children	Home swimming pools
Daily Mail	Woman	Spotting property hot spots
The Independent	Minimalist interior	Fashionable London interiors

Look at the list of the main pictures — not an exterior shot of a house or flat among them, and a million miles from the old-style estate agency-style photographs. Even *The Independent,* which was one of just two property supplements not to feature a person or people in the main cover picture, used a lifestyle shot of an über-fashionable interior rather than a "traditional" picture of a property exterior.

The conclusion from all this provides the answer to the question: "What does a journalist look for when writing for a national newspaper?".

Put simply, stories about people and lifestyles are almost universally popular with readers of all the national newspapers be they broadsheets, compacts or tabloids; stories simply about homes without any human involvement are not particularly popular and do not sell newspapers.

If you delve further into an analysis of this particular week's property supplements you see even more evidence of the importance of people and lifestyle. Let's look in detail at the *Sunday Times'* property section Home. It has a 1.3m circulation and because of that high readership and its authoritative status it is considered by advertising chiefs and PRs as the most desirable paper in which to get a mention of a property on sale or a new development being built.

The vast majority of Home's stories had high "people" content, even if they were supposedly about one specific property or even a largely statistical subject, such as the state of the housing market.

Lifestyle stories:

- Author Joanna Briscoe on family life in Devon longhouse (people).
- Jonathan Ross moving house (people).
- Stirling Moss as landlord (people).
- Ozzy Osbourne selling up (people).
- London playright house-hunting (people).
- Living next door to someone with mental illness, with case study (people).
- Archers' actress gardening tips (people).

Pure property stories:

- Norfolk former air control tower for conversion into homes (not people).
- House of the week (not people).
- Conservation area restrictions with two case studies (people).
- Easton Neston estate finally sold, interview with buyers (people).
- Hassles while building extension, with case study (people).
- Hampstead properties (not people).
- Spanish homes in the Ebro Valley, with case studies (people).
- Glass walls used in houses (not people).
- Comparative prices of castles on sale (not people).
- Homes for sale in Spain (not people).
- How do you build a home from scratch? Story (not people).
- Faults in a new home (not people).

Market stories:

- Homes taking a year to sell, with three case studies of sellers (people).
- Tales of a landlady with case study about renters (people).
- Market snippets (not people).

- Should you sell now or wait? (not people).

From this we can see that of the 23 stories in Home, 13 had people and their experiences at the heart of the content, even if the stories' main thrusts were about a specific property or the state of the market.

The final piece of this analysis concerns one particular story in that edition of Home, which specifically looked at properties taking a year to sell. The individual properties faltered on the market because of a dramatic fall in the number of transactions during the first half of 2005, successive interest rate rises in the preceding year, and what some onlookers believed to be the start of a wider economic slowdown at that time.

At first sight, the story appeared almost statistical — not surprising because it is, after all, a market story. It contained a table of average property prices in June 2004, December 2004 and June 2005 to show the slight fall in prices over time, and it included details of how the number of sales had dropped by 30% during that period.

But the story comes to life because of a trio of case studies that show the human effect of this slowdown.

A London businessman was relieved that his house eventually sold after 12 months, although only as a result of his slashing £225,000 from its price. A Cotswolds art consultant was frustrated because she had changed estate agent twice and axed £255,000 from her house, yet it was still on the market without a buyer. The third case study was celebrity model Jodie Kidd who had withdrawn her property from the market because it could not attract a buyer at a price she stuck to over the 12-month period.

Each case study involved specific prices and times, quotes from most of the people involved, and photographs of them. Readers could feel genuine sympathy for the frustration of the individuals, and the scale of their dilemmas and losses were clear. Estate agents, buying agents and market analysts were quoted too, giving the story a combination of powerful human interest and strong authority.

This was, in construction, a typical and successful story for a national newspaper property section.

Case study: Analysis of Estates Gazette coverage

Contrast this now with an analysis of a very different sector of the press — property industry publications and specifically *Estates Gazette* (EG).

The 3 September 2005 issue of EG had 144 pages. When advertisements, the editor's personal page and pre-assigned columnists' articles had been removed, there remained 46 pages of stories where property industry players and their rival PR representatives could fight for attention.

Pure property stories:

- Seven pages of property news (32 brief stories in total).

- Five-page story on Wiggins Group/Planestation.
- Two-page story on commercial and residential franchising.
- Focus: Lancashire and Cumbria business parks.
- Focus: Lancashire and Cumbria industrial sites.
- Focus: South Wales shopping centres.
- Focus: South Wales industrial sheds.

Market stories:

- Two pages of market and finance news (11 brief stories in total).
- One page of property share analysis.
- Focus: Lancashire and Cumbria retail market.
- Focus: South Wales office market.
- Marketplace: seven brief stories.

Lifestyle stories:

- Energy efficiency and how it can be used in commercial property.
- Property life gossip column (seven short stories).
- Events (four short stories).
- People (20 short stories).

Even in this heavy-duty industry publication, there is a strong "people" tone in the coverage of stories. To show this, look at the breakdown of pictures that were used across all of the editorial pages in this edition:

- Pictures and images featuring people: 25.
- Pictures and images featuring properties: 27.

Then study the nuances of many of the stories — they may have been primarily industry stories, but many featured people at the heart of the content. For example, much of the Wiggins Group/Planestation piece was about Oliver Iny, a flamboyant character behind the company; in many other editions of EG there have been profiles of leading individuals in companies or sectors.

"Pitching" commercial property stories

While press releases about deals will always provide a flow of information to fill the news pages of the commercial property press, there are countless other opportunities to identify and "pitch" feature stories to the commercial property media.

Research reports — Most of the large surveying practices invest significant sums in their research departments, who produce an incredible stream of original and useful research about issues as diverse as office rentals across international markets, yield values in prime retail stock, remaining available brownfield sites and out-of-town retail development. This research is produced primarily to attract business clients. However, sadly many firms miss a trick by failing to involve their PR advisers in the research study at the outset — to identify what angles will be of interest to different types of media, how the research might be modified to increase media interest or even how the tables should be presented and made available to encourage media use. There is usually little communication, even when the research report is finally made available to the press — the research reports often focus on the facts and fail to include sufficiently media-friendly comment or opinion or translate the material into practical advice or clear messages from a named "authoritative" individual for different audiences.

Profiles — Returning to the human interest element, most magazines will be happy to run a profile of a senior or interesting figure in the company if they are prepared to make interesting comments, give an unusual view of a topical subject or be photographed pursuing their particular out-of-work interest. While many senior people feel disinclined to become the focus of a major piece, they often fail to recognise that a great deal of good material can be conveyed about their company and its developments and products along the way — so it is not so much personal PR as corporate PR with a human interest angle.

Technical developments — Most property magazines (and many business and management magazines) will allocate space for short updates and explanations of new legal and other regulatory developments. Alerting features editors to these developments — and their impact on or relevance to their readership — is usually sufficient to gain a commission for the piece. The byline is usually given to the author — sometimes supported with a statement of their expertise or experience in an area — and usually includes a mention of the author's organisation. If you have senior people within your company who are also able and willing to provide good-quality copy on a regular basis you might even persuade a magazine to allow the individual to become a regular contributor or even a columnist on a particular area of expertise. A regular property slot in a leading business or management medium can provide a useful source of sales leads as well as a higher profile among the readership.

Round tables — Another proven method is to do all the leg work in organising a round table for a magazine. A topical or important issue is identified, between six and 12 interesting individuals are approached and invited to participate. Some of these individuals will be sufficiently senior or well known (or their organisations will be well known) to have a broad appeal. The round table can then be "pitched" to a magazine with an invitation for them to send one of their staff or freelance writers to attend so that they can write up the discussion reflecting the specific interests of that magazine and its readership. This approach has been successfully used to gain coverage of a new emerging business locality, a pressing development issue and imminent new legislative and its likely impact.

The common elements in these approaches are: the research and lateral thinking to identify a topic that would be relevant to the target readership, having the right individuals available to take ownership of the issue and talk authoritatively and interestingly.

We have made the point — people count when it comes to winning PR. Having seen the distinct types of outlet and how they treat their stories, let us now turn our attention to press releases and how they should be strictly tailored to suit the target outlet.

Press releases: Good and bad examples

A vivid example of how good press releases work and bad ones do not, came in the immediate aftermath of the announcement on 6 July 2005 that the 2012 Olympic Games would be staged in London.

There had already been substantial speculation in the mainstream property press about how the London market generally and the east London market specifically would benefit from the regeneration schemes required to prepare for the games. But now it was time for that to be set out in more detail.

With the announcement date declared some five months in advance, there was no excuse for being ill-prepared — yet many estate agents and developers appeared to be caught on the hop, as were their publicists. The best prepared participants got the best results, as will no doubt be the case at the real Olympics too.

For example, within an hour of the announcement, the Halifax issued a detailed, 1200-word press release that was to become the definitive comment on the subject from any part of the property industry.

It started like this:

House prices go for gold in Olympic host cities

As London celebrates being awarded the 2012 summer Olympic Games, Halifax has examined the link between hosting an Olympic Games and the impact on house prices.

The regeneration effects from hosting an Olympic Games has generally had a positive impact on house prices. Each of the previous four host cities have seen house prices rise by more than the national average over the five-year period in the run-up to the Olympic Games, the main period of Olympic-related development activity. The level of outperformance was, on average, 18 percentage points.

Barcelona was the best performer with prices rising by 131% versus an 83% increase in Spanish house prices in the five years leading up to the 1992 Olympics.

Table 1 House Price changes in Olympic cities in the 5 yrs leading up to the Games

	5 yr % increase host city	5 yr % increase host nation	Difference
1992 Barcelona	131%	83%	49%
1996 Atlanta	19%	13%	7%
2000 Sydney	50%	39%	11%
2004 Athens*	63%	55%	8%
Average	66%	47%	18%

*Data for Greece is for 4.75 years not 5 yrs

Key Findings:

- Hosting an Olympics is usually associated not only with an increase in sporting facilities but also an upgrade of transport and cultural/leisure facilities. Barcelona, Athens and Sydney all saw a significant upgrading of their urban infrastructure and this city rejuvenation is likely to encourage higher house prices.
- In Barcelona, amongst the Olympic related projects, 78km of new roads were created. There was a 17% increase in the sewage system, a 78% increase in green zones & beaches and a 268% increase in the number of ponds & fountains.
- The scale of Olympic redevelopment is reflected in the high costs associated with hosting an Olympics. The Barcelona Olympics was the most costly of the recent Games, at an estimated cost £8.1bn. The estimated costs for other Olympics were for Atlanta £1.5bn, Sydney £2.5bn and Athens £8.0bn.

Tim Crawford, Group Economist at Halifax, said:

"This is great news for London. Hosting an Olympic Games encourages city regeneration and is usually accompanied by an improvement in facilities and transport links. These factors tend to be positive for house prices. Homeowners in Hackney and Stratford could potentially reap similar benefits to other Olympic precincts over the longer term."

The rest of the release used well-researched statistical evidence to examine the good and bad effects of the hosting of the Olympics on the housing markets in recent host cities, and how the 2002 Commonwealth Games aided the Manchester housing market.

To help humanise the statistics, case studies were provided by Halifax estate agents to help television, radio and newspapers find "real people" who were going to move to east London because of the likely property bonanza they may enjoy.

Within a further hour the release had been used on the BBC News and *The Guardian* websites and it became an ingredient of the many hours of coverage given to the Games announcement that evening on BBC, ITN and CNN news, and on BBC Radio and Independent Local Radio news.

It also featured repeatedly in the scores of pages of coverage in the following morning's newspapers including *The Times*, *The Guardian*, *The Independent* and the *Financial Times*.

What was so right about this release? Essentially, it gave an oven-ready story that was rich in facts and evidence. In detail that meant:

- There was no need to ring anyone for a quote, because one was included.
- The research was clearly well-prepared and hinged around the obvious domestic angle (London house prices) by looking at what effect previous Games had on other comparable cities.
- The table was made available in downloadable form that could be cut and pasted into websites or newspaper make-up pages.
- It was made available very quickly.
- Crucially, it was backed up by case studies willing to be quoted and photographed.

Compare that with the slower, apparently off-the-cuff comments from many other property industry players.

For example, one of Britain's leading top-end estate agents issued this 94-word statement some hours after the International Olympic Committee announcement:

> [Company name] congratulates London on its successful Olympics bid to host the 2012 Olympic Games and looks forward to the potential spin-off benefits for London's housing market.
>
> The anticipated economic activity and infrastructure improvements should provide a timely uplift to one of London's largest regeneration areas. At the same time, the anticipated delivery of in excess of 9,000 new housing units will, in one fell swoop, provide 15% of London Thames Gateway's target housing requirement between 2003–2016 (60,000 units according to the ODPM) and at least 50% of the 9,000 units will be affordable.

What was wrong with this? Well, it was very short (not a problem in itself), but it failed to provide any evidence or informed speculation about what the Olympics might mean in detail to an area or to house prices, and the only statistics issued were those repeatedly cited in stories some months earlier about London housing targets. It also started off "[Company name] congratulates ... ," which made it sound like a dull corporate release and not a punchy "This is what it means for London and home owners" statement.

It should be said that this was a rare exception from this particular firm, which has commercial and residential press offices that are normally exemplary and regarded by journalists as being among the very best in the industry. But a week later the firm confirmed that this particular release had not been "picked up" (journalists' slang for "published") by any outlet.

A smaller but high-profile rival residential agent appeared even less prepared, issuing a 285-word statement from four different spokesmen in the company (one in charge of the Exeter office, an area with no direct involvement in the Olympics). The only figure used in it was "2012" and its national press comment led off, rather eccentrically, with the predicted beneficial effects of the Olympics on East Anglia:

> For those of us who need to travel east of London it will be the catalyst for the re-building of the road and rail systems which in the long run are the arteries to East Anglia. The result will be to the benefit of all towns and cities in the East Anglian region.

Once again, this release was not picked up by any outlet.

Most bizarre of all was a press release issued on behalf of a small east London estate agency on Monday 11 July. This was five days after the Olympics announcement and one day after many Sunday papers' news sections and property supplements had run pieces on the effects of the Games on east London's housing market.

The release was to have gone out on the previous Thursday (the day after the announcement) but had been held over to avoid getting overlooked in the coverage of that day's terrorist bombings in London. Had it been released on the day of the announcement (as had the Halifax release), it could have achieved much more publicity, it would not have become lost in the welter of coverage about the bombing, and it would have been of great help to journalists seeking a different view on what the Games would mean to property prices.

As it was, the release was not picked up by any outlet, although it could have been if more research had been used in its content and if more emphasis had been given to the second and third paragraphs. These contained a more interesting (and probably more truthful) perspective than those parroted out in releases from many other agents. The problem was, it came far too late:

[Estate agent] of [name] estate agents comments, "London hosting the Olympics in 2012 obviously has implications for the area in which most of the build will be taking place — ie, East London. If the planners have a wider, long-term vision and consider the existing infrastructure of the area, the Games will be a huge boost to communities, in that they will bring with them more facilities and improve things in what were hitherto fairly under-invested in parts, as happened for Manchester at the Commonwealth Games.

However, it will have very little immediate impact on property prices in the location; the increase will be more long term as the areas become more accessible and, consequently, more gentrified. At the moment, Our office in Hackney is nearer, as the crow flies, to the City than is our office on Clerkenwell, and the arrival of the Olympics will see transport links improve to the extent that areas like Dalston and the Kingsland Basin might become the milieu of choice for the well-heeled stockbroker.

A note of caution should be sounded amid the lift in morale — East London residents are, on the whole, great users of their local amenities, and places like Hackney Marshes, which at the weekend sees lots of people out flying kites and playing football, must be protected and not have big stadia built over the top.

While some can say the Olympics announcement was a one-off, it was nevertheless a perfect example of how the well-prepared, well-researched release can make a huge impact.

The same principles apply to much less high-profile issues and announcements. For example, here are two press releases about the property market that misfire because they did not adhere to the golden rules about timeliness and human interest:

12 July 2005: [COMPANY NAME] PROPERTY REPORT HIGHLIGHTS A RECOVERY IN PRIME RESIDENTIAL PROPERTY VALUES

Property advisor [company name] has launched its annual report on the prime central London market. According to [company's] prime property value Index, property prices in central London rose by an estimated 1.75% in 2004. This is on the back of no growth in capital values during 2003 and supports the view that the central London market has seen a marginal recovery in 2004.

While the volume of sales in England and Wales declined in 2004, both London and the South East saw transactions increase over the same period by 3.78% and 3.57% respectively. Activity in London has focused on the new homes market and the very top-end of the established homes market.

In 2004, purchases of new homes in the prime central London market have predominantly originated from UK although consortium-led investors from Ireland have been noticeable. In the established homes market, foreign buyers are displaying a particular interest in £3 million-plus properties. However, a large proportion of these overseas buyers tend to reside in the UK for part, if not all, of the year. While Russian, Middle Eastern and European buyers have been influential, Sterling's strength against the US dollar has deterred US and Far East investors.

[Agent's name], [company's] residential managing director, comments, "In summary, the outlook is positive. Although the UK economy is expected to slow in 2005, London is in a much better position. The labour market remains strong and households are benefiting from growth in earnings. In addition, interest rates remain low and contrary to what the pessimists might say, we do not expect a collapse in the housing market.

What was wrong with this? It may have had a quote and the full report may have been rich in genuine research but it had one crucial weakness — it was released in July 2005 but it referred to 2004, and it was as "dry" as desert sand.

Literally dozens of property consultancies, financial analysts and estate agents had poured out their statistics and review of the 2004 market by the end of the first quarter of 2005. So when the external public relations team hired by the agency tried to offer this report as an exclusive to the *Sunday Times* midway through the third quarter of the following year, it failed to make the cut. They also did not bother to find a case study to help humanise it, despite the paper's reputation for advocating such an approach.

Here is another market press release that misfired, this time from the north of England:

10 August 2005
[Agent's name], Senior Associate Director of Residential Lettings and Management Services at [Company's name] responds to current media reports on a booming buy-to-let market with a view on the impact this is having at local level:

"The buy-to-let market is one of the great success stories of the last 10 years and investors have reaped the financial rewards. However, it is a cyclical market that is subject to the laws of supply and demand and following a slower start to the year the buy-to-let market in Yorkshire is presently in a more stable phase.

"We have seen a shift in activity over the past couple of months with the majority of people entering the market looking to make longer term investments. They view property as an attractive alternative to investment in pension or stocks and shares.

"Potential investors are increasingly researching the area and market in more detail than before, with one and two bedroom apartments remaining a popular choice amongst the buy-to-let market, especially if they are in close proximity to road and railway links — these assets being particularly important to tenants.

"With the Bank of England's decision to cut interest rates to 4.5% and suggestion within the media this week that further rate cuts may be on the horizon we believe there will be an up-turn in new enquiries. As long as prospective investors continue to recognise property acquisition as a medium to long term investment, the buy-to-let market will remain very solid and continue to be a popular option."

This release really is all at sea.

First, it is difficult to see the point of the story — if you were writing a headline for it, what would it concentrate on? Would it say that buy-to-let was a 10-year success? Would it say that there has been a "shift in activity" (but from what to what)? Might it hint that property investment remains a popular option (but compared with what)?

Second, it is steeped in generalisations with absolutely no objective evidence whatsoever to support any comment. The firm could have given numerical examples of this "shift in activity", it could have set out how purchase prices and rents in the buy-to-let sector had changed, or given statistics about numbers of enquiries. But it didn't.

The impression is that the firm either had little to say or that the market was bad and this was being concealed by vague comment; in reality, neither was the case because the firm is a well-respected player in the north of England. But the release succeeds in giving a false and unfavourable impression.

This press release may have had the odd paragraph that could slot into a local newspaper but it was sent to national journalists and, unsurprisingly, not one used it.

But there are many examples of highly successful releases about market information. Some of the best come each month from the PR machine working at Rightmove, the residential sales and rental website. Its monthly analysis of asking prices provides a rich source of stories. This release in early August 2005 is typical:

SUMMER SALE AS REALITY STRIKES AND SELLERS DROP PRICES BY 1%

The Rightmove House Price Index for July shows that asking prices fell by an unexpectedly large 1% this month, the biggest monthly fall since November 2004. This has resulted in a further drop in the annual rate of increase from 2.4% to 0.2% (a meagre £451), which is the lowest for 10 years.

Sellers are finally willing to help bridge the affordability gap to attract greater volumes of buyers, mimicking the retail High Street by cutting prices in the fashion of a 'summer sale'. There are positive signals for both buyers and sellers as increases in real wages combined with now-static house prices over the last year, mean that house prices are now more affordable than their peak last year. With an anticipated cut in interest rates and signs of recovering demand the prospect of a price crash is further reduced.

More information in the attached report.

That report includes many pages of regional and national statistics, quotes from an economist, and detailed breakdowns of information that are perfect for local and national newspapers alike. Unsurprisingly, Rightmove received coverage in almost every national newspaper, BBC News Online, property and financial magazines in return for a thorough job well done, and produced literally within two days of the data being gathered.

Rightmove, which has its own in-house PR team, also helpfully issues this data each month with a four-day embargo, allowing newspapers to find, interview and photograph case studies in good time for the agreed release of the material.

The same good and bad examples can be seen when estate agents and PRs seek to publicise individual properties or developments. With newspapers and broadcasters demanding human interest angles, it can be difficult to get media interest for something as potentially dull as a block of flats.

This press release does not even try:

AUGUST 2005: FOR IMMEDIATE RELEASE PHOENIX RISING

On the site of the former [name] Public House, like a phoenix rising from the ashes, [development name], a contemporary development of 26 apartments, is taking shape.

With 30% of the apartments already sold off plan, [development name], with its fabulous seascape views, is certain to be a landmark property for the future.

The development is being marketed jointly by [developer's name] and [estate agent's name]. Both agents are delighted by the level of interest shown to date, but not surprised because they recognise that this contemporary scheme, with stylish interiors, in such a good location as this will always be popular. With views to over 180°, with Felixstowe to the left, sweeping round to Walton and the Naze on the right, demand was always going to be high.

[Agent's name] of [estate agency's name] said "Irrespective of economic commentary, when location and property are right, buyers will always be found. The development certainly ticks all the boxes; with a view that is constantly changing with cruise liners, ferries and container ships coming and going and small yachts always on the horizon, together with the apartments proximity to the amenities of [location name]. The development will suit all buyers; those seeking somewhere to retire, those downsizing and also those seeking a stylish first home.

A comprehensive brochure is available from [agent's name] and [developer's name] and interested parties should make contact with either agent to avoid disappointment.

[Estate agent's name] continued "once the show apartment is available to be viewed later this year, we are certain that the remaining apartments will be sold very quickly".

Bizarrely, there is no reference to price, there is no picture to accompany it and, although you get an impression of the view from the properties for sale, you have no idea whatsoever about the design — is it a tower? are there flats or duplexes? do they all have a view of the water?

Jeremy Dodd, a senior PR at Barratt Howe agency, explains that there are different kinds of press release and some of the worst are those that appear to placate a client rather than convince a journalist.

There are still [client] companies that demand a news release per month, per development to prove a level of 'productivity'. Their PR companies are inevitably scraping around looking for anything that moves to draft a news release on it. This explains the steady flood of 'Move in to Great Oaks for £99 this weekend' and 'Experienced Annabella joins sales team at Nether Wotsit' news releases

claims Dodd.

The downside for the PR industry is that there are a number of PR consultancies, staffed by junior people, who literally churn out news releases with very little imagination for clients who are process-led rather than marketing-led. Because the results are poor, client satisfaction is poor, the fee levels are low, consultancies cannot afford to raise the quality of the staff, which means the results remain poor. It is a self-perpetuating downward spiral.

However, it could be very different if the PR can persuade the client to produce the material that journalists actually want.

Here is a press release about potentially uninteresting properties but it has been turned into an opportunistic dream — and there is nothing wrong with opportunism if it simultaneously meets journalists' needs and promotes a client's products.

28 August 2005:
Cricket Mania Knocks Oval Apartments for Six

Cricket fans are offering £23,500 a week to rent an apartment with a terrace looking directly into The Oval during The Ashes in September.

Eight newly refurbished duplex apartments with views directly over The Oval's pitch are available to rent for the week of The Ashes through leading London estate agent Kinleigh Folkard & Hayward. Offers have already come in for sums of £17,500 followed closely by one for £23,500.

"This is incredible," comments estate agent Kinleigh Folkard & Hayward's director, Richard Forshaw. "Apartments like this would normally fetch a rental value of approximately £500 per week, so these figures are quite exceptional.

"On the other hand, this may well be the most important test match in 20 years for England, and as there are no other apartments with outside space offering views like this, you can't put conventional values on them. This also makes it an interesting investment opportunity for purchasers when The Ashes come back to England in four years' time."

The eight split-level apartments are currently on the market for sale in Kennington Oval priced from £360,000 to £495,000. At the top of a Victorian mansion block, built in circa 1880, each apartment has been refurbished to a high standard and comprises two bedrooms with two en-suite bathrooms, a further WC, dining room, kitchen reception room and decked terrace with amazing views across The Oval cricket ground.

The release was e-mailed simultaneously to the news desks of newspapers as well as the property sections and specialist property journalists — and it paid off.

Because of the interest in The Ashes test series taking place at the time, many newspapers ran the story on the news or sport pages, giving enormous publicity to a small estate agent and its apartments in what is, normally, an unexciting and low-profile area of London.

The release was accompanied by pictures of the flats, which were used by many papers.

Good pictures are vital

The importance of good pictures to accompany a story may be obvious but the reality is very different. Indeed, the absence of high-quality pictures has fast become a major stumbling block that prevents many otherwise-excellent property stories appearing in print.

"We need better images — nothing cheesy or with phoney models. Brush up and improve computer graphic images (CGIs) and make sure all electronic images are high-resolution at 300dpi. If PRs don't understand this, they should learn," says the editor of one consumer-oriented magazine who believes photographs help to widen its appeal to a wider audience than just "property enthusiasts".

Once again, we can see why this subject is important by analysing an example.

On 17 July 2005 the *Sunday Times'* Home supplement contained 56 pages. Discounting those pages wholly or partly containing advertisements, its journalistic editorial content included 73 photographs broken down like this:

- 21 photographs, primarily of people.
- 23 photographs, primarily of specific properties.
- 11 photographs, primarily of gardens.
- 12 photographs of pieces of specific furniture (not general property interiors).
- 9 photographs of general views of areas.

For the reasons discussed earlier, it should come as no surprise that there were 73 photographs but only 23 stories, because multiple pictures or photo-montages are now common to make pages stand out, grab attention and "humanise" stories. Nor is it surprising that properties and gardens combined constituted only 35 photographs out of 73, or less than 50%.

The reason comes back to the editorial direction of the section, and for that matter most other property sections — the huge emphasis on lifestyle and people.

This is conveyed more vividly in photographs than in text, and more vividly in pictures of people with "real" stories to tell than with stock shots of properties (shots that may have been doctored by estate agents or developers to ensure a backdrop of blue skies and a foreground of affluent youthful residents).

It is crucial for PRs, developers and estate agents to remember that a good story as told to the journalist over the telephone may sound perfect but it may never be used if there is not a photograph to illustrate it. Sometimes this means that the PR should ensure a good story comes with a good picture, ready to supply to the newspaper.

There are problems of course, especially if the PR's client is a small estate agency or developer who has never seen the justification in using professional photography.

"Clients with digital cameras can be difficult. After all, we can all take snaps of properties and e-mail them around these days and it costs next to nothing," says Tim Stanley, a Bristol-based PR consultant who specialises in work for commercial and residential developers and agents in the UK and overseas.

"The PR has a big persuasion role here and it demands a degree of tact if the enthusiastic snapper is to be convinced. A good picture does not happen by chance. It has to tell the story and must be taken on a professional camera to produce jpegs of a specification which meets the demanding criteria of picture editors," says Stanley.

"The client may be proud of his photo of the new building, despite the refuse skip and discarded cardboard boxes in the foreground, but it fragments when it is enlarged on the computer screen. The professional's shot will hide the rubbish, make the building look even more appealing, illuminate the dark corner, and enlarge to fill half a page in a glossy magazine, if required," he says.

Indeed, a professional photograph often takes on a life of its own, beyond the article for which it was taken.

It is little known among the general public but outside of the news, sport, business and magazine sections, most national newspapers rely on either library pictures they may have had in stock for years or outside "donated" pictures to illustrate whole sections of newspapers.

Travel, motoring, personal finance and property sections normally have tight budgets and if unique pictures are ever commissioned from professional photographers acting for the newspaper, they are more likely to be of human case studies than of existing houses or new developments. The same applies, only more so, to most magazines where there are few in-house editorial staff so freelance contributors are relied upon heavily and there is no budget for "unique" photographs.

Therefore, if a PR can get good quality pictures to accompany a press release, it is much more likely to be picked up than a release without any picture at all. Indeed, really good pictures may be used occasionally for years to come to help illustrate other articles.

But if those pictures do not come through, it can be the death knell to even the best story as told to the best journalist.

For example, the editor of a residential property magazine wrote to contributors in July 2005 to say:

> We have got to get more and better images. Please read carefully the commissioning letter sent to you — it has the details of what pics are required. We don't like rejecting features — especially if they're good — but if we continue to run features about property accompanied by pictures of rolling fields, beaches and donkeys we will lose readers to the competition. So fair warning, features will be rejected without the requisite number of good images. Please check sizes as well as composition before sending them through. You are required to supply six property photographs — either in print form or electronically at a minimum of 300dpi and 10cm × 10cm in size — all with relevant captions on a separate Word document. If captions aren't supplied in this format then your article will be rejected.

This particular editor may be risking goodwill among contributors by threatening to reject stories on the basis of photographic quality but his point is valid. Readers want to see what they are reading about, and the people involved too, and they want good quality shots that have been professionally photographed. What goes for one relatively small-circulation magazine goes, in spades, for a national newspaper.

There is one other important point here that a good PR will take on board, and that is to vary the pictures supplied from paper to paper and from magazine to magazine.

Another magazine editor puts it this way: "The biggest problem that we have with property PRs is that everything is sent out *en masse*. We never seem to have opportunities for an exclusive, or even just a different angle. Images are standard — the very same brief selection seems to be sent to all magazines — and I really don't want to be publishing the same photos as my competition."

Estate agents, developers and their public relations representatives sometimes forget this. There is one incentive not to do so in future — if you don't do it, your rival probably will.

Perhaps the best known example of a PR agency getting this right is Robert Barlow Communications (RBC), which adopts a shrewd and successful approach to its work on behalf of Barratt Homes.

Barratt is a respected housebuilder but, inevitably, its products are regarded as homogeneous and visually unexciting to property editors wanting unusual properties with startling images. So instead of sending out pictures of similar-looking new homes, RBC sends out pictures of the homes' buyers with accompanying press releases that emphasise the human interest rather than the predictable descriptions of new apartments and houses.

So to promote a rather plain-looking Barratt development called East Central in the east London borough of Hackney, RBC issued a picture of David Smith, a 51-year-old Hackney council manager who had just bought a flat there. The property in question was in the background but the main photographic subject was the buyer, and the details of the press release explained how the development was well-located for the buyer's work and weekend interests.

Likewise, to promote a development at Thamesmead in the Thames Gateway, RBC sent out a photograph of a buy-to-let buyer called Shirley Amedzro, with a press release stating how she had bought a property at the site as an investment.

RBC's pictures are taken by professional photographers and are frequently used by national newspapers, lifestyle magazines and trade publications. The subject matter — human beings — and the high-quality photography gives Barratt a far greater number of mentions in stories than would ever be the case if the company relied solely on pictures of its properties. The same can be done to "popularise" relatively dry corporate activities in the commercial sector too.

Garvis Snook, chief executive of Rok property solutions, says: "Our brand values incorporate our mould-breaking approach to many things — including having fun. For our results one year, my finance director and myself dressed in leathers and used an electric guitar and Harley Davidson motorcycle for the photos. The *Financial Times* produced the photo full size. Some of our investors were a little negative but internally it was massive — a real example of our commitment to being a fun, exciting, challenging and different company. This example is now used in business schools to show how to communicate visually what a brand really represents" (see p119).

The message is simple: complement a story with good-quality, high-definition photographs, 300dpi and 10cm × 10cm in size, showing properties and people, to give your product a much better chance of being used in publications.

The importance of case studies

Today, case studies are the lifeblood of national and local newspapers, lifestyle magazines and increasingly trade publications too. They all want to link stories with their readers, to make people identify with or aspire to the activities of the people and properties featured.

So in a national newspaper a story about how there is a move towards city-centre living is only viable if it includes examples with people willing to say why they moved to city centres, to speak of the pros and cons of their new home and location, and to say what they paid for their property — and, of course, to be photographed too.

Likewise, a lifestyle magazine looking at retirement accommodation will only carry the story if there are case studies of buyers. Even trade magazines carrying stories about different types of industrial units often require occupiers to give the low-down on the new premises.

Property professionals often say they are reticent to ask clients about the slightly more personal reasons about why they sell, buy or rent. As a result, they find it difficult or impossible to obtain case studies. Perhaps some estate agents genuinely are reticent but for those who are serious about obtaining a publicity profile, this is now a crucial part of the exercise.

The *Mail on Sunday*'s property supplement, which is circulated only in the Home Counties, runs a weekly column about first-time buyers who reveal their relative poverty and struggle to get on the property ladder. This may sound a tough job for a journalist tasked with finding new examples each week but, in reality, many people are willing to give even personal financial details in return for their 15 minutes of fame in a newspaper.

We have seen from the Barratt Homes photograph examples that this approach really works, so claiming shyness about speaking with clients may lead to them losing out to rivals.

Commercial property is, at the best of times, a much harder sell to newspapers, magazines or anything else other trade publications. But a case study may help here, too, as Stuart Sapcote, managing director of Sapcote Developments explains:

> Commercial and industrial is fairly hard going to get a journalist interested in, I've found. It's a bit difficult trying to interest someone in a letting of a 50,000ft shed — the whole story has rather a cold feel to it. If, however, one can add in a human interest story — for instance the tenant or occupant has an unusual trade, or is the biggest operator of his type in the industry — then suddenly there is a story that might be of interest.

Exclusives

Exclusives tend to be associated with breaking stories in Sunday tabloids about celebrity indiscretions, but they can sometimes apply to property journalism too.

Occasionally, a major story — it may concern an important grand house, a home owned by a celebrity, or a development attracting unusually high-profile interest — is given to a newspaper on an exclusive basis by a PR representing an estate agency, property consultancy or developer.

For a journalist and an editor, "exclusive" means being given to that one publication or broadcast programme only, with the promise that it will not be revealed to any other outlet in any medium until it has been used by the title to which it has been given.

When this approach works, it works well. The story is usually so big that it receives special treatment with plenty of space, pictures and favourable coverage.

Simple enough? Perhaps — but many disputes have arisen in the use of exclusives.

Journalist Cheryl Markosky says such disputes are perhaps more likely to happen because of the sheer number of property sections that exist in newspapers, magazines and even on websites.

> With extra competition in the field, the 'bad' PRs promise clients coverage everywhere. This hacks off decent national journalists who give exclusives to their newspaper of choice, and trust breaks down rapidly when that promised exclusive ends up in more than one paper. I find that if I don't ask if the story has run before and if it is exclusive to me — surely, I shouldn't have to do this? — I can come unstuck.

Newspapers do not always help PRs, their clients, nor even the journalists when it comes to exclusives.

Sometimes a journalist will file an exclusive promptly but the newspaper will sit on it for some weeks, perhaps because of a reduction in the number of pages available for editorial. This, in turn, leads the PR firm that offered the exclusive (or its client owning the property in question) to become frustrated. They may then agree to give the story to a rival outlet, annoying the original recipient of a story that is now no longer genuinely exclusive.

Sometimes a PR will be over-ambitious and give two different angles of the same story to different publications, without mentioning one to the other and so making each believe it has a truly "exclusive" story. Only when both stories are published on perhaps the same day do the rival titles know they have been misled.

On other occasions, stories that are unbelievably dull are sexed-up by PRs who offer them as an "exclusive" to a journalist or newspaper in the hope that this tag will suggest the stories have more worth than they really deserve.

In any of these circumstances, when an exclusive goes wrong or an exclusive tag is deliberately misplayed by PRs, there is only one long-term conclusion — the PRs and their agent or developer clients become less trusted and respected by journalists.

This is not to discourage the use of exclusives, but simply to say that there is only one definition of exclusive that is recognised by an editor or journalist. This means the story is theirs, and only theirs, until they use it. If a PR cannot stick with that, then do not describe the story as exclusive. If a PR misplays the exclusive game, then the property firm engaging that PR should look around for someone new.

The combination of the need for exclusives and the domination of property journalism by freelance writers have effectively forced the demise of the "property launch", that gathering of journalists to mark the debut of a new development or a specific house that comes to the market.

> In the early days of property PR most of the journalists were permanently employed by the publications; now most of the journalists are freelances. To these journalists, time is money. To go to endless launches is a waste of their time when they can receive the information by e-mail and talk to the agent and get information on the telephone. The last time we gave a launch party, against our better judgment, one journalist turned up

says Margie Coldrey, a veteran of property PR for many years.

Her agency, MCPR, handles the PR for national agency Jackson-Stops & Staff and upper-end local agents such as Marchand Petit.

Means of communication

For editors, staff writers and freelance journalists alike, it is crucially important that PRs make contact with the right sort of information, at the right time, and in the right format.

It is not the job of estate agents or developers to know the idiosyncrasies of different publications and individual journalists — that is why it is sensible for property professionals to employ PRs in the first place — but there are some common rules that apply to almost every journalist, irrespective of title, status or ability.

Use e-mail as the main form of communication

"I prefer to be e-mailed than rung up unless it is really urgent," explains Carey Scott, editor of the Home section on the *Sunday Times*. "I ask for e-mail not post — I get 45 letters a day and I'm a one-man band," says Nigel Lewis, editor of the *Daily Mail*'s Friday property section.

"Any cold call that comes through our reception, I ask to go straight through to voicemail" is how another editor handles the plethora of PR approaches to him.

Most journalists communicate between themselves and their editors by e-mail; most also request information from PRs by e-mail. This latter communication can take the form of round robins to PRs but increasingly sophisticated online systems are emerging for journalists and PRs to use.

One of the newest is *www.responsesource.com*, which is rapidly gaining favour among journalists.

They complete a form on the homepage of the journalist section of the site. It is automatically sent to PRs representing the property industry who have subscribed to the service. If the PRs can help, they reply directly to the journalist by the deadline entered on the original enquiry.

The site encourages journalists to seek facts, comment, background information, interviews, products to review, statistics, photos, case studies and news — the only "out of bounds" uses are seeking or offering sponsorship or advertising. Once the enquiry has been distributed, usually within 15 minutes of being sent by the journalist, it is up to the PRs to contact the journalist directly, usually by e-mail.

There is one other major bonus for journalists — for them, the service is free.

Public relations agencies or the PR wings of estate agents or developers have to pay to use the service, but again the rewards are considerable.

First, any subscribing PR puts his agency on the list of recipients for the journalists' enquiries outlined above. A PR firm has to register in one or more "sectors" on the vast Response Source database, which contains thousands of PR firms with clients straddling hundreds of disciplines.

The most appropriate sectors for property PRs are Business & Finance (which covers property industry activities), Public Sector & Legal (everything from planning issues to social housing), the obvious Construction & Property, Personal Finance (looking at mortgages, investment properties and so on), and Home & Garden (covering everything from interior design to gardening tips).

PRs going the extra mile for their journalist contacts and their clients may also want to register with sectors of the database that may occasionally be contacted by non-property journalists. For example, some property PRs register with the Women's Interest sector to help provide case studies of female buyers and renters to those publications that write human interest stories.

PRs using Response Source also have access to a special online wire service on which they can post their press releases (costing £30–100 per release, depending on scale of circulation required to those journalists and newspapers registered with the service).

The PRs can also access a freelance journalist contact list compiled automatically by the website as individual journalists register. Finally, the PRs can also set up special advertising pages to publicise their own services — or those of their clients — when journalists make contact, at variable fees up to £450 plus VAT.

Delivery of promised stories

If I say no to a story it's because I have decided it won't work for us, and I am unlikely to change my mind so there isn't much point flogging a dead horse and arguing with me about it. PRs who give you 'facts' that later turn out to be wrong are a nuisance, as are PRs that promise something or someone that they can't deliver

says the *Sunday Times'* Carey Scott.

This is a common complaint and is one focus of tension between journalists and PRs.

Roger Hunt, a freelance contributor to a range of trade publications and with a particular interest in period property, says the gap between PRs' promises and delivery can often be great but is sometimes because of a lack of support from PRs' clients.

> A huge number of PRs fail to deliver. They enthusiastically promise the earth and simply fail to come back with any information. A key point for their clients to remember is that while they may spend good money on PR they must realise they have to support their agency, otherwise it is left having to apologise to the journalist for not being able to get the information

he explains.

"There are certain PRs and house builders that I refuse to deal with because they have repeatedly let me down in the past by making promises and then not delivering or because they have been rude and aggressive," says Hunt, whose views are typical of most freelance writers.

However, Melanie Bien, one of the very few to successfully transfer from journalism to public relations within the property world, says that although PRs have a tough job, journalists have a tougher one when it comes to delivery.

"PRs won't thank me for this but I think the pressures on a journalist are greater. The buck ultimately stops with you as a journalist; if a PR can't find you a case study, for example, that's it as far as they are concerned. But a journalist has to get one from somewhere so you can't just leave it at that," says Bien, who used to be a section editor on *The Independent* but now handles PR for Savills Private Finance.

Learning the lessons

So what are the lessons to learn from these analyses? There are crucial ones:

- Always tailor press releases to ensure they provide the right sort of information and content to get in your chosen outlet(s).
- The more information you provide the better because journalists — especially freelances — work on many stories at the same time and are happy to have some of the pressure removed by someone doing research for them.

- If you are hoping for local or national newspaper exposure, ideally provide a case study that illustrates the point — ensure the case study is willing to talk about prices, their experiences buying/selling/renting, and they must be happy to be photographed if a newspaper or magazine wishes.
- If you want national newspapers to take your views on the market seriously, include impartial, well-researched information and not just anecdotal views or cynically misinterpreted figures. Why not commission an independent external source or undertake research in-house? Either way, it must be seen to be thorough and authoritative.
- Make sure comment is timely — often the first well-considered press release on a subject helps set the agenda for many subsequent stories, so waiting a day or more to agree wording may make a release ineffective.
- Never be afraid to tell it as it is. Estate agents that honestly say prices are falling or that some properties are hard to sell in certain circumstances will invariably receive more publicity and be trusted more than those who only co-operate with journalists when the market is rising.
- Do not base a publicity campaign on "launch ceremonies", as they are often attended only by the least-busy journalists who do not write for major publications. Instead, target publicity at specific publications and choose journalists that write for those titles. As we will see later in this book, journalists and editors cherish exclusives. The publicity from targeting, say, four premier-league titles may be greater than a scattergun launch attracting second-division attendees.
- If you promise an exclusive, honour that promise.

But there is one overarching lesson that encompasses all of the above and more.

Freelance journalist Catherine Moye, who writes for the *Sunday Telegraph, Mail on Sunday* and *Financial Times* property pages, says property professionals and PRs must think "outside the box" and must have complete knowledge of their properties and, more importantly, the stories behind them.

She tells a vivid and true story of how PRs can missell a story and almost lose vital publicity for a property client.

I get a release saying an Italian development is set in 40 acres surrounded by olive trees and vineyards — frankly, where isn't in Tuscany?. Then there are lengthy descriptions saying it is great for investors. The two *Telegraphs* [Saturday and Sunday, both of which have property sections] turn it down on the basis of 'well all very nice, but so what?'

Moye herself visited the property in her own time while on holiday, not expecting to get a story from it because of the indifferent information supplied by the PR firm.

"It turns out that the Scottish developer has been coming to Tuscany all his life. He chances upon this historic village, decides to do it up into a swanky five-star hotel and apartments for sale, and brings over a whole Glasgow team who learn to speak Italian," says Moye, who believes that even cursory research would have shown that this was a good human interest story for a newspaper.

The hotel chef was a former employee of Gordon Ramsey and the restaurant manager worked at a restaurant in Glasgow when JK Rowling sat in there day in, day out, writing Harry Potter. All these Scots hold an Italian Burns night in Tuscany when the Italian staff don kilts and even played the accordian. But the best I get out of the PRs is that the hotel presses its own olive oil from olive trees on the estate. Wow, hold the front page!

So the developer lost at least a full page of publicity in the *Daily* or *Sunday Telegraph* because the PR failed to research the story properly and gave cursory details to a journalist. Yet, by sheer luck, the same journalist found out what a good story it really would have been.

How often does this happen, when neither the PR's client nor the journalist ever realises?

There is a clear lesson here. Whether you are a PR or a client, look for the most interesting, unusual angles even if they are not directly about the property you are selling. Or even if they are not about property at all.

Reactive PR: Direct Contact with Journalists

This chapter is about a conundrum.

Residential and commercial property agents, developers, surveyors and architects deal routinely with literally millions of pounds worth of assets, make pitches to unforgiving clients and then sell to the public. Some have day-to-day responsibility for scores or hundreds of contractors, craftsmen, architects, suppliers, sales staff and visitors, and have budgets running into hundreds of millions. They may frequently have to answer to demanding shareholders or business partners keen to know about, and get some reward from, the financial success of their projects.

Yet many of these same powerful individuals find it hard to speak directly with journalists. Why?

Partly this may be because of an unjustified suspicion that a journalist will, by definition, try to catch people out. In reality, this is not true on 99% of occasions because of the non-contentious nature of most stories.

Partly, it may also be because some property professionals now hire PR companies, so believe direct contact with journalists is consequently unnecessary. Again, this is not true — the PRs' job is to ensure that such contacts are managed and show the individual and company in the most favourable light as part of a broader marketing strategy.

If direct contact is unavoidable and inevitable, property professionals deserve to have an insight into how the journalists work and what they want from the interview.

Much direct contact will be in the company of PRs, as we have seen in chapter 2.

Because most property stories are predictable and uncontroversial, many journalists agree to have interviews on the telephone or in person "lined up" through PRs one or two hours (or more likely one or two days) in advance. The gap between a journalist's request and the interview allows a PR to test the water with the journalist, find out the story's angle and tone, perhaps even obtain a list of questions or topics to be discussed.

The PR should pass this information on to the client; a good PR will also give an honest assessment of the journalist to the client. Is he or she a "soft" writer, specialising in articles about style, architecture, furnishings and interiors? Or will they be writing a more probing article about the financial state of the company, the health of its turnover or the vagaries of the sector in question?

Have recent articles by the journalist and/or in the publication in question been supportive or hostile to the activities of the person or company being represented at the interview?

This is bread and butter to experienced property PRs and should be part of their service to clients. Journalists know this, even if they do not always like it.

Reasons why journalists may contact you directly

There are some interview circumstances where a PR is not present if, for example, a journalist rings an estate agent or developer out of the blue. On other occasions, even if a PR minder is in the room or around the lunch table, he or she will not be the person having to answer the journalist's questions.

This chapter is a guide to how and why this happens, and gives some tips on how to manage the situation to achieve the best possible outcome in publicity terms and maintain good relationships with the media.

The tension between journalism and PR

There is a historic (some say inevitable) tension between journalists and PRs. In an ideal world it would not exist but it does, in property and all other areas of editorial coverage.

Precisely why this is so is buried in mythology and past practice. Perhaps it is that PRs tend to want to express only the positive image of a client while a journalist tends to want to accentuate the negative — after all, "Nasty developer builds shoddy house for poor elderly buyer" makes for a more interesting story than "99% of homes are fault-free" even though both statements may be true.

Perhaps also it is because many PR agencies, and in-house press teams, tend to have extremely high levels of churn — staff leave, move to rival firms, or drop out of the PR industry in high numbers. Such professional promiscuity may discourage loyalty and support from the journalists who deal with them.

On a more subjective level, property PR also has an image of being staffed primarily by under-30s, while the majority of commercial and residential property writers are over-40s.

Whatever the reason, the reality is that the relationship can be fraught. This is absolutely not a justification for anyone in the property industry to consider it better to do without public relations — journalists may be a complaining, dissatisfied breed but they never say PRs should not exist, merely that they should be more responsive.

The tension between journalism and PR works in both directions and takes many forms.

> I'm scared to contact the *Sunday Times* property section because the team there know that they run the publication every PR wants to get their client into. As a result, the paper doesn't reply to me if it's not interested in a story. I understand it has taken ideas from PRs and given it to their favoured writers without referring to the clients who inspired it

is a complaint from one PR.

> It seems unnecessary on the journalists' part to be quite so abrasive or not to have the decency to respond at all. I've never understood why someone can't take 20 seconds to reply to an e-mail. I've run a large newspaper section single-handedly and always had plenty of time to answer e-mails, check voice messages and still feel at times that I didn't have enough to do

according to Zoe Dare Hall, a freelance property writer, who used to be a staff editor at a newspaper.

Journalists, of course, complain in the other direction.

The results of a survey of 40 of Britain's most prolific residential property writers for this book found that all but one regarded most residential PRs as providing inaccurate material, not being sensitive to deadlines or printing schedules when delivering information or contacting journalists.

But no journalist suggested he or she could live without PRs completely. The tone, from more than one writer, was simple: "I can't live with them but I certainly can't live without them. Property PR is here to stay."

Reasons why journalists sometimes do not use PRs

After reading of the tensions between the two, it may be tempting to assume journalists and PRs try to avoid one another like the plague. That is untrue, although there are occasions when a journalist will go directly to a company's staff without contacting the firm's PR representatives first.

One practical reason for this is that newspaper, radio, television or online deadlines can be very tight and it is not always practical to "line up" interviews in advance.

For example, the *Sunday Times'* Home section is written by freelance and staff journalists and the property editor during the week before press day, which is Wednesday each week. The prepared section is then reviewed by the overall editor of the newspaper, John Witherow, at about 2.30pm on Wednesday afternoon. The section, if agreed, is printed very early on Thursday morning rather than later in the week, in order to leave time and energy for preparing and printing the more time-sensitive Sport, Business and News sections, which are reviewed and printed much closer to the Saturday evening cut-off point. Late on Saturday evening, the completed newspaper is then distributed around the country and overseas.

If Witherow dislikes a story in the Home section on Wednesday afternoon, it is spiked and another needs to be selected to fill its place within about 90 minutes so that it can be sub-edited and re-checked prior to sign-off and then printing.

This may mean that a journalist may have to write something from scratch or check the facts of a standby story that may have been prepared up to three months beforehand and will need rapid updating.

This happened in early August 2005 when Witherow took a dislike to a story about John Prescott's proposals to streamline town planning procedures. The story had been researched and written, and photographs had been taken to illustrate it. Then came the news that the editor felt too much had been written about Prescott in the previous week's issue — even though those stories had not been about planning.

The author of the Prescott article was contacted and another piece he had written some seven weeks earlier was substituted. But this alternative story (which was about the old standby of "what do you get for £1m in the current property market") contained references to specific properties and had interviews with buyers and sellers that had been researched a long time before. Seven weeks on, there were clear concerns about the story — were those properties still on sale? Had prices been reduced or increased to reflect market changes in the intervening period? Were the interviewees who were described as buyers and sellers still in those roles now?

There were only 90 minutes to check the information, so the writer did not contact the PRs who had originally helped provide the property details and case studies. The PRs may have been in meetings, working on other stories with other clients, or may even have been on holiday.

Instead, the journalist went directly to the estate agents, developers and individual case studies to check that the quotes, prices and details still held good almost two months later. When one of the original quoted agents was unavailable, the frantic journalist dropped this reference and rang another agent — again without going through a PR — to find new information.

This is typical of how most newspapers operate. For example, other nationals' print times include Monday evening (for the Wednesday Property section of *The Independent*), Thursday afternoon (for the Bricks & Mortar supplement of *The Times* on Friday, and the *Daily Telegraph*'s Saturday Property section), Friday afternoon (for *Sunday Telegraph*'s House and Home) and Friday evening (for the Cash supplement, containing property stories, in *The Observer*).

There are other, and less obvious, reasons why a journalist may deliberately avoid going through a PR even if, on this occasion, he or she has time to consult them.

The reasons for this include:

- The existence of a good professional relationship between a journalist and an agent or developer that means they can speak together directly without a PR setting up the interview (often PRs are aware of this in principle and are happy about it because it saves their time).
- A journalist may want to obtain access to someone who is normally "shielded" by a PR. Most journalists know of key people in an organisation (and this applies to all organisations, not just property ones) who will provide outspoken views, which may often be more interesting than the bland, on the record, carefully manicured comments from "official spokesmen". This is not to

say the outspoken individual is necessarily wrong or not telling the truth — often he or she might be giving a completely realistic picture, just as the line of the PR is to give a different, less truthful story that may reflect a better image on the firm. Either way, the PR will want to shield him or her to prevent potential problems for the company, and a journalist will simply have to ignore the PR to make contact.

- A journalist wanting to pretend he or she is not a journalist in order to get information that would not be made available through official routes. This happens routinely in investigative journalism, whether it is the high-profile *News of the World* type "stings" with a journalist claiming to be a contact for a drug dealer in order to expose the wrong-doer, or whether it is in the less melodramatic world of property. Incidents like this happen rarely in property but when they do occur they are controversial, so let us expand on this.

Mystery shopping

Journalists writing on the residential property market frequently go "mystery shopping" to try to judge the state of the sales and rentals market. This is done without revealing their identity as writers. The journalists simply call estate agents or new developments' sales suites that are close to the action, pretending to be members of the public.

For example, the journalist may claim to be an individual buyer trying to strike a deal on a property that might have been lingering on the market for a year. Or he or she may claim to be a well-informed landlord wanting to bulk-purchase half a dozen buy-to-let flats in a development where sales are rumoured to have been faltering. This is exactly how investigative market articles, frequently found in most property sections of national newspapers, have to be written. Almost all serious property journalists have done this on numerous occasions.

Some newspapers, notably the *Sunday Times* and *Mail on Sunday*, make a virtue of saying how they "covertly" found this sort of information; they make it clear that their journalists went undercover in order to express to readers that these papers are on their side, championing the consumer and finding the bargains on their behalf. This is a common philosophy of most newspapers, be they broadsheet, compact or tabloid.

A few estate agents and developers believe such coverage is good; after all, it saves them spending thousands of pounds taking out an advertisement to say they are offering cut-price properties, or extra fixtures and fittings, or other incentives.

Other agents and developers regard this covert journalism as underhand and believe the publicity given to a weak market merely adds to the decline of confidence and causes further weakness. Very occasionally, these property companies' PR representatives will even rebuke journalists for using underhand tactics.

This is an unwise move as such criticism is the proverbial water off a duck's back for most journalists who are immune to criticism of a story unless it shows the story to be inaccurate. "Tough luck" is the polite version of a typical journalist's response.

In a consumer-oriented society we all enjoy reading where to get the best deal on the consumables around us, everything from mobile telephones to new cars, and property is no different. Declining to introduce oneself as a journalist when making a telephone call to a sales suite may be sharp practice to a few but it is difficult to see how journalists could operate any other way and still get the facts of the story.

Worse still for property professionals, there is almost no way of identifying such approaches.

Having said all this, it is important to remember the context. Very few calls from journalists are covert and most deal with straightforward, non-contentious matters. When covert calls do occur, they are normally about prices and occur when the market is on the turn, downwards.

The rest of the time, journalists and property professionals will have wholly transparent and straightforward dealings. So how should you manage them?

Handling enquiries from national media property journalists

In the national press arena, it is often the case that an agent or developer will be tipped off in advance that a journalist will call. Property journalists frequently tell PRs of the stories they are writing, to see if appropriate experts or properties can be found to help build the stories.

If Mr X in a particular property consultancy is the right person for the job, and that is known by the PR, it is usually the case that he will be asked to ring the journalist at an agreed time, or vice versa.

When direct contact occurs, following this checklist will help the property professional get as much from the interview as the journalist:

- *Confirm the publication* (because many journalists are freelances, it may not be obvious when they identify themselves by name).
- Then check the *subject of the story* and do not be afraid to ask the approach being taken by the journalist — for example, if an article is about penthouses on new developments, it could be about their flamboyant design and image of wealthy bachelor buyers, or it could be about how they are so much more expensive than similar-sized properties on lower floors. Always check.
- Ask what the *deadline* is by which the journalist needs information from you (it may be immediate or often up to a week later — if it is the latter, you may have time to find out more evidence to support your views).
- Never be afraid to *ask if other agents* are being spoken to for the article, and if so, who? In theory, this may be of academic interest but it is common for three to 10 interviews to take place for a major article in a newspaper.
- Give *simple answers* appropriate to the publication's readership (for example, an interview for a story about solar panels used in *The Independent* will have to simplify technical issues; an interview for *Estates Gazette* about remediating soil on brownfield sites can, and will, be much more detailed).
- Remember journalists like *facts* as much as, or more than, opinions. If the story is about the state of the property market, genuine figures about business volumes and sales prices versus asking prices will impress the journalist and give the interviewee and his or her organisation a much better chance of being included in the story. A set of generalisations that mean very little are a switch-off to good journalists.
- Try not to use property *jargon* that can be misconstrued. Estate agents use the term "price correction" when they mean "price fall" and suggest vendors should be "flexible" when they mean they should "cut prices" — journalists are many things but they are not fools, so they see through jargon.
- If an interviewee needs to *get back to a journalist* with information, he or she should do so quickly, even if the journalist's deadline is not so urgent. This is because with so much information coming in from a variety of property sources, any journalist may find just the right ingredient from another source while waiting for the original interviewee to return a call.
- Try to find *case studies* to support the interviewee's position and the thrust of the article. Remember that newspapers, magazines

and even trade publications are increasingly keen to make property stories more human.

- *Do not ask* to see the article before it is printed, even if it is just to "check for accuracy". The inevitable refusal by any good journalist may appear to be classic editorial arrogance but it is actually very practical — if 10 people are interviewed for an article, getting their agreement on its content would never be achievable within a reasonable time frame. In any case, a sub-editor at the newspaper may alter the story after it is filed to fit the space available and to complement the photographs used, so any agreement on precise wording between interviewer and interviewee will be null and void before publication.
- *After the interview*, the interviewee should deliver a brief call or short e-mail to the company's PR representative to let him or her know how it went. A good PR will e-mail the journalist both to get a feel for how the interview went and to offer any additional points or information that was missed during the interview.

The culture of national newspapers tends to be sceptical and fact-based. This does not mean that the property industry will be given a rough ride (remember, property journalists want continuing good links with industry insiders, so they rarely burn their bridges).

But that culture does mean that if an estate agent is asserting that the property market is in good shape and that prices are edging up in a sustainable fashion, he will be asked to back that up with evidence; if a developer says his homes are more energy-efficient than his rivals', then he must be able to prove that too.

In reality, few interviews of this kind are ever anything but agreeable, amicable and mutually useful. Few estate agents, surveyors, architects or developers on top of their subject would find it difficult to deliver short, sharp, accurate answers with evidence to back them up.

It may sound absurd, but having taken all of this into account it is also important to ensure you "say something", that is that you avoid being so cautious and inoffensive that your quotes become bland and not worth using.

Simon Slater of First Counsel, a management consultancy in the professional services market, was formerly head of marketing and corporate communications with DTZ Holdings and director of marketing with Eversheds.

He says:

If you don't stand for something, you are perceived to stand for nothing. Ensure you or your organisation always has 'a view'. Develop a platform of beliefs about your market and forge a reputation as an influential player. But take a long-term view of reputation and always remind those around you that whilst it can take 10 years to build a good reputation, that same reputation can be destroyed in just 10 days.

There is one final point — and it is a delicate one. When giving an interview do you play a straight bat and tell is as it is, or do you talk up the market or your product in a shameless way?

Christopher Stoakes, an author and former editor of specialist publications including *International Financial Law Review*, *The Tax Journal*, *Global Investor* and *Risk Financier*, says too many professionals fudge an answer to avoid appearing negative or to avoid giving away "free" advice.

"In fact readers see right through that. I urge the opposite: give the answer and clients will come to you. Frankly, if it's simple enough to be answered in an article, you don't want that sort of work anyway because you can't charge for it," he says.

Handling interviews with friendly journalists on local newspapers

Just as the culture of national journalism is sceptical, so the culture of most local papers is almost entirely the reverse. The local press does not want to publish untruths but this genre sees its primary role as that of uncritical "cheerleader" for all activities taking place on its patch — and that includes the health of the property market.

Look, for example, at one of Britain's most successful daily regional papers, the *Western Morning News*; its general news section is riddled with stories claiming that West Country food, beaches, lifestyle, tourism and even farming traditions are better than those elsewhere in the UK and overseas.

Its property section, published as an insert to the main newspaper each Saturday under the title Westcountry Homes, follows a similarly uncritical path.

Its 27 August 2005 edition was typical:

- A front page filled by a photograph of a restored farmhouse in what appeared like an editorial story but was, in fact, an advertisement paid for by the local office of a national country house estate agency, which was marketing the property in question.

- Although many indicators at the time were suggesting a poorly performing property market, the Westcountry Homes main editorial story spoke of a "surge of first-time buyers in the region" helping to boost the market — a generous interpretation of vague figures produced by agents who were also advertisers.
- Of the 19 other pieces of editorial in the edition, all were about individual properties on sale by advertisers and all were gushing in their praise — headlines like "Stunning Old Mill on Banks of River" and "Live The Good Life in Comfort" give the clear impression that this is unchallenging journalism.

So it is with the journalists who contribute to Westcountry Homes and most other local newspapers' property supplements. Usually, they are part-time journalists towards the end of their careers (with a small number of honourable exceptions, of course) and they are discouraged from being critical because local estate agents and developers spend so much money on advertising.

"[Local newspaper] Editors are super-sensitive when it comes to estate agents — a notoriously difficult bunch whose revenue is, of course, so valued," says the lead writer of one local newspaper property supplement.

Anything critical of the property industry tends to be avoided, as the same journalist found out when she tried to include an article that called for estate agency practices to be modernised. "It was pulled at the last minute by those more senior than I who feared that it may cause too many problems by offending some estate agents," she says.

This attitude is evident in interviews too, so for the estate agent, developer, surveyor or architect who is asked to be an interviewee, there is no reason not to take full advantage:

- As with nationals, *confirm* the publication in question and subject of the story, and the deadline by which the journalist needs information to be supplied.
- Most professionals could *offer to write the article* themselves if the newspaper would accept it — most local property sections include directly written pieces from estate agents in particular, especially on the state of the market or on the characteristics of key localities in the circulation footprint of the paper.

- Be prepared to *offer photographs* of properties that are for sale to bolster the article — local papers are notoriously reliant on "given" images to illustrate articles as their budgets for photographers will be tight.
- On this occasion, an interviewee *can ask to see* an article before it is published or, at the very least, ask to see any quotes from, or references to, his or her organisation contained within the piece.

For many serious journalists and indeed many serious estate agents, local newspapers are "good things" but entirely unchallenging.

"We're in our local newspaper because it would look odd if we weren't but, to be honest, it doesn't really help us sell properties," says the head of one of the largest regional offices of a leading country house estate agency. Much the same is said by his rival agents too, suggesting that there is a two-tier approach to local newspapers.

The first is that for middle and low-market estate agents, local papers are vital organs for publicising the properties on sale and the brand of the agency itself. The second is that for large agents operating at the top end of the market, most local newspapers outside of the six largest cities in the UK are useful primarily for building the brand, but not for selling the properties themselves.

Handling interviews with unfriendly journalists on a national or local paper

Picture this. You are a Glasgow estate agent and you pick up the telephone on a sleepy Monday morning to hear the caller identify himself as a reporter from the city's *Evening Times*.

"The owner of a house that's on the market with your firm says it's impossible to sell because of the local council's plans to open a centre for drug addicts two streets away. Is the local drugs problem stopping other properties selling too?," he asks.

Now what do you do?

This is minute-to-minute fodder for local and national newspapers alike and, although property professionals will rarely find themselves in the front line of such a story, it is advisable to be prepared as best you can.

If the story really is untrue and you feel you can say that with authority, do so. Local newspapers in particular are prone to run stories based on one irate citizen and if you know that view to be cranky or unrepresentative, say so. But often the one irate citizen is merely the first of many to hold that view, so be careful not to set yourself or your company against a strongly held view.

So let us assume that while you are aware of the drug problem in general, in reality it is not a major factor for the sale of that particular property or others in the locality that are being handled by your company.

You need to get this across in a rational way that minimises damage to the individual seller, and to others nearby, and to your business more generally.

Here are a few hints of what *not* to do:

- *Do not assume you can break* into the interview and ring your PR for help. He or she may not be at work early in the morning and the journalist does not have time to wait for a discussion (most evening papers have their first deadlines in the early morning).
- Do not try to appeal to the journalist's corporate instinct by saying how much *advertising* your firm places with his or her paper. Unlike a dedicated property journalist, he does not need to keep on your good side; even if it is a local paper, the news desk will be much more bullish towards you than a compliant property writer. In any case, he or she knows front-page banner headlines about irate house sellers help both circulations and journalistic careers.
- *Do not put down the telephone* or brusquely say you will ring the reporter back and then choose not to do it. This is, effectively, a confirmation of the story to the reporter, because you chose not to deny it.
- Do not say *"no comment"* if you really mean to deny the story because "no comment" also, effectively, confirms a story to a journalist. In our example, any good journalist would hear the term "no comment" and then write something along the lines of: "A representative of the estate agency refused to deny that the seller's house was effectively unsellable because of the drugs problem".

Instead, try to get points across calmly, intelligently and logically.

In this case, you may suggest that there are drugs problems in many areas of large cities and their property markets are not necessarily affected; or that even areas with tough reputations such as Glasgow have seen house prices soar in recent years and that highly localised factors do not always come into play even in so-called "difficult" areas.

When the call concludes you should alert your PR representative to get a more considered response for later editions of the paper; the PR will also help you prepare in case another local publication or a regional television or radio station "picks up" the story and develops it further, later in the day. A good PR will then find out from you what the real problem is, and will brief you on the best way to respond to it when further enquiries are made.

But that sober, rational, moderately worded response is required at first unless you want your hostility, or your silence, or your "no comment" comment to be interpreted negatively.

Handling enquiries on a negative story

There are occasions when the property media have to investigate and report on a story that you would prefer they did not. In the commercial world, this could be when the media hear rumours of problems on a major project, that a firm is experiencing serious financial problems, that redundancies are imminent, when key people are about to leave, about disaffected clients, a less than favourable outcome from a professional negligence court case or when news of a possible merger hits the streets. In some cases, there may be an upset employee who leaks information deliberately or accidentally.

It is important to remember that not all media attention at a "crisis" is unwelcome. There have been occasions where totally inaccurate rumours can be nipped in the bud by early and positive media relations intervention. There are also situations where merger or dispute negotiations have been speeded up or resolved to avoid the impact of negative publicity.

The journalist will know that in these cases sometimes the PRs are unaware of the story themselves and that if they are they will do their utmost to stall or deter the journalist while maintaining their established good relationship. Therefore, the journalist will try to find out what is happening by either calling a number of their existing contacts within the organisation or even cold calling senior and junior people.

When an organisation becomes aware of this, they should ideally put in place their media relations crisis plan. Sadly, most organisations do not have such a plan until they have experienced the trauma and negative publicity that can occur when a planned and co-ordinated approach has not been available.

As always, honesty is the best policy. If — for legal or commercially sensitive reasons — you are unable to provide information, then you should say so. Recognise that the journalist has a duty to follow up their lead and report on the story and find a way to help them in a different way or at some later date — when the information is not so restricted.

Regardless of how terrible a potential story is, you must recognise that being evasive or rude to journalists is only going to make the situation worse. You must be as co-operative and as open as possible if you are to have a hope of assisting your side of the story being conveyed.

A good crisis management plan will have the following components:

- *Early and strong communications internally.* This should involve the senior management team, the team whose "crisis" it is and the PRs. Too often, bad situations are made worse when the PRs are kept in the dark, subsequently misinform the journalists (destroying credibility and sometimes long-established relationships) and are cut out from managing the ongoing situation.

- *Analyse the story and develop a strategy.* All the relevant information should be identified and collected — and all the players (which could include site staff, third parties, lawyers etc) consulted. PRs should advise on the worst and best possible outcomes so that everyone's expectations are realistic. A co-ordinated approach should then be agreed.

- *Nominate senior spokespeople.* Agree who should take calls and ensure that the nominated spokespeople are senior enough to talk credibly about the situation and trained in dealing with the media under difficult circumstances.

- *Prepare a short, factual statement about the situation* for release and ensure it is communicated to those media who need to now. Ensure that the contact points are identified and that these people are available on a 24-hour basis.

- *Communicate the situation internally.* Ensure that anyone who is dealing with or likely to be approached by the media has a copy of the statement and knows to where they should direct any media enquiries. Everyone should also advise the central PRs if they have any media interest in the matter. If the key spokespeople have been given a "for internal use only" list of questions and answers to ensure they provide accurate, up-to-date and consistent information to the media, it may be worth providing this information to others in the organisation who may be contacted by the media.

- If the crisis relates to a *particular location* (rather than the organisation itself), make sure that a senior person of the organisation goes to the location immediately. This shows that the organisation acknowledges the severity of the situation and is prepared to do whatever it can to resolve any problems.

- If possible, avoid situations where several reporters are there together — it is much harder to control things when this happens. An alternative is to provide *one-to-one briefings* with journalists so that each can pursue their particular line of enquiry without the other journalists all jumping on the bandwagon.

- If you have some particularly *strong relationships with the media,* now is the time to call upon those journalists and offer them greater access to the senior people and information than you might to journalists with whom you have had little prior contact. By concentrating your efforts on those journalists who know your organisation already, at least they are more likely to give you a fair hearing and put your side of things across — thus minimising the negative impact.

Regardless of how things progress on the crisis, it is important that after the story has ceased to be of interest that the PRs follow up all the journalists involved and tries to help them with other stories and re-establish good relations.

Occasionally, a PR or property player with a good, ongoing relationship with a journalist can use their professional relationship as leverage.

For example, Garvis Snook, chief executive of Rok, says that on one occasion his firm was working as part of a consortium on a major project covered by a confidentiality agreement. At a separate press briefing, one of his directors inadvertently referred to the project in response to an unrelated question.

> The journalist from a south western daily newspaper jumped on the reference — it was breaking news of a major project with massive interest to his readers. Luckily, there were a number of prior dealings with this journalist and we explained that a mistake had been made and how breaking the story at that time could jeopardise the project. He agreed to cover the piece in a low-key way at the time in exchange for an equally interesting story in the region

explains Snook.

There is one other piece of advice on the aftermath of negative stories, should they reach print, broadcast or a website. Do not overreact.

This case study from an in-house press officer at a surveying practice shows what can happen:

> The worst situation is when a journalist or a magazine gets it wrong and the partners or directors overreact. The situation is compounded when the upset partners or directors do not understand how the media works and adopt an angry and retaliatory stance. There was one situation where — as a result of a bitter legal dispute — there was an unfortunate photograph taken of a senior person. Naturally, the individual was incensed at the insensitivity of using such an image — particularly alongside some rather negative press comment which contained one or two small inaccuracies. In this situation, lawyers became involved and writs were issued. Whilst the directors felt they were justified in taking some action, the impact on the editor and journalists involved was obviously very severe. The result was a lose-lose situation all around and very little chance of ever building positive relations with the magazine again.

Handling local or national television

If dealing with an aggressive news desk press journalist is rare, so having to deal with a television or radio interview is rarer still — we may think property is an irresistible subject but, in reality, it is not a major story to most broadcasters.

If you are approached by a broadcaster, the likelihood is that it will be about a local market story. For example, have first-time buyers plummeted in number? Or is Acacia Avenue about to see its first home on sale for £1m? And is the average price of a home in your patch falling after several years of rises?

Here are key points to remember about television interviews:

- The interviews take a long time to set up so you will get *substantial notice*. A producer or researcher will ring you, usually a day or more in advance, to seek your company's participation.
- Ask your PR, if you have one, to *prepare a mock interview* to give you or your chosen spokesperson a rehearsal — a good PR will also be able to hire a journalist or producer with television experience who will be able to give expert advice if the interview is important enough to merit the effort and cost.
- A "*package*" is the bundle of information that makes up a report; it consists of the introductory "cue" material read by a newsreader, the reporter's voiceover and pictures giving the substance of the

story, the interview with one or more people, and then often a concluding "piece to camera" by the reporter before handing back to the studio. Few packages will last longer than two minutes, so the interview section will be very, very short — so manage your expectations.

- Do not let political correctness get in the way of projecting your company's best image — put forward a *spokesperson* who is articulate, smart, photogenic and authoritative even if that person is not the most senior or the oldest.
- Most local news television interviews are edited in advance, with few answers allowed to run for more than *20 seconds* — the more answers your spokesperson delivers inside this time, the more likely they will be used.
- Television is much more superficial than written journalism and you can get only *three key messages* across at most in 20 seconds — prepare them and, unlike with newspapers, do not clutter them with too many statistics. Just say: "The market's sliding because there are 10% fewer first-time buyers than last year and there's twice the number of homes on sale. That's why prices are dipping a little," for example, and do not try to remember and splutter out many more facts and statistics.
- As always, *get the opinion of a PR* after the edited interview has been transmitted.

Managing radio interviews

Although television steals the limelight, radio can be much more challenging. Local radio stations will frequently ask estate agents to be the interviewee or guest on phone-in programmes, or even do live telephone interviews on to news programmes.

Again, there is sensible preparation to be done to help you get the best out of such an opportunity:

- Confirm the *radio programme and station*, who your interviewer will be, how long you will be on air, whether it is live or recorded, and at what time it is broadcast.
- *Establish the subject* of the interview and *prepare* a few basic facts in writing (and ensure they are on card and not paper to avoid it rustling near a microphone or if you are shaking with nerves).
- *Never try to read a statement* or key points out verbatim — this will

sound stilted and unnatural. Try to be natural and conversational and, as with television, make your answers brief but authoritative.

- Make sure that if the programme is recorded, your PR goes through your performance with you. It may well be that you will be able to *issue a press release* based on what you said and receive additional newspaper publicity.

Direct contact with journalists via e-mail

A growing phenomenon in property journalism is the "e-interview" or "e-statement". For most journalists, this is not a worthy substitute for a telephone or face-to-face interview because it is almost impossible to ask "spontaneous" questions by e-mail.

But with increasing pressures on time and more editorial space given to overseas property, e-communications is a good way of interviewing across time zones and — even if it is an exchange of e-mails between people who are just a few miles from each other — this method can give real advantages to the property professional. For example:

- Any response by e-mail gives the opportunity for the agent or developer to *carefully consider* the words used, and to prepare some research if appropriate.
- The answers or statements can be *checked with the PR first* before sending to the interviewer.
- There is the practical point that prompt e-mail exchange is *quicker* than random telephone calls between busy people who may be out or unable to comment for long.
- The spokesperson can keep an *exact record* of what was said, to be checked against the published story.

Some PR agencies put a lot of effort into getting written comments on issues for journalists — as with any story, this could be exclusive (JagoDean, a London PR agency, pioneered this approach in the late-1990s) or it could be a general press release to all journalists that was sent out via e-mail instead of by slower means.

Now many journalists, too, are happy with e-mailed comments. Major stories with contentious angles still need live interviews (allowing the interviewer and interviewee to make more spontaneous comments, triggered by earlier answers or questions) but the bulk of routine stories can, in theory, be handled by e-mail.

Sometimes whole stories are made from passing comments prepared by agents or developers and e-mailed to journalists — on the one hand this could be described as lazy journalism, but on the other hand it is a good way of getting a story written rapidly with more authoritative comment than could be achieved by "hit and miss" telephone calls and then bedevilled by disputes over alleged misquotes.

Multiple outlets for an interview

Whichever way you give an interview, be prepared for it being used possibly more than once.

Freelance journalists in particular have to make interviews or visits "sweat" to justify their time, so it is possible that your words on, say, the explosion of new apartments to buy in Bulgaria will appear in more than one article that a freelance writes. As we have seen, fees for magazines — in particular property-related ones with relatively small circulations — are small, so it is common enough for articles to be reversioned.

It is also possible that a newspaper may syndicate a story; for example, in the UK, *The Independent* syndicates some stories to the *Irish Independent* and *The Guardian* has deals with sister publications in North America. European editions of UK newspapers may also carry articles. Likewise, there may be crossover between media. *Estates Gazette* pieces are sometimes carried on *www.egi.com* for example.

Of course, this will not apply to every story, but be prepared for your interview to reappear.

Handling journalists on press trips

A press trip is a physical visit by one or more journalists to a property, development or location in a bid to get a better understanding, to spawn story ideas, and to help create a continuing working relationship.

Press trips are pro-active rather than reactive but they are included here because, unlike a press release or press launch, much of the homework is done by the agent or developer rather than the PR.

It is he or she who will have to set up contacts or possibly case studies in a foreign country if it is an international press trip, and it is he or she who will physically lead the tour in most cases — talking through with a journalist the specifics of a property, a development or a location.

Joy Moon, who has led marketing and public relations for up-market estate agent Strutt & Parker since the 1980s, says: "When I started my career you spoke to a journalist about a property, arranged a visit and then left finding the story to them. I even remember the days of taking up to 20 journalists to the same property together."

But things have changed. "The key now lies not with the client but with the agent who needs to understand what makes a story and seek out information at the start of the instruction and is able then to brief the PR office properly and manage client expectations," says Moon.

Unsurprisingly, most property professionals would rather have their teeth extracted without anaesthetic than spend substantial time with journalists in this way ... well, perhaps. But a well-handled press trip can be hugely productive for journalists and property professionals alike.

The good news for property people is that PRs usually initiate the idea, checking with them about the appropriateness of the one or more journalist invited.

Not so long ago, these exercises used to be real junkets — extended trips where the property or development was a minor part of what was otherwise a substantial freebie for the journalist. There remains stories about journalists who blagged their wives or husbands along, or who were so enamoured with the local vino that they were never able to remember the details of the property when the trip was over.

As freelances have come to dominate much property writing, so the press trip has come of age as a useful tool. Quite simply, any freelance who is away from his or her PC and is not writing stories is also not earning money, so these days press trips are much more focused.

The key to success or failure is whether the principles outlined in chapter 3 (exclusivity, good pictures, access to case studies and the like) are adhered to in the trip. To ensure they are, there must be clear and prepared thinking by the property professionals.

For example, in 2005 Graham Norwood, one of the authors of this book, went on a well-organised press trip where every factor for his story was taken care of by the PR and the developer. Good use was made of everyone's time, an exclusive story was presented to him, and the result was substantial volumes of positive publicity for the client.

Story: first "buy-to-let" properties in Moscow being marketed at British buyers
Preparation: journalist given a wholly exclusive story by a PR, with time to pitch it to editors. *The Times'* Bricks & Mortar section agreed to take it. Minimum amount of time agreed for trip would be 48 hours including London-Moscow return flights.

Trip details: Meet at airport with PR and the London representative of the developer; discussed the development, Moscow property market and related issues en route to Moscow. The itinerary at the destination included:

- Visit by car and on foot to two developments constructed by the developer.
- Meeting with two developers and interpreter, with developers leading discussion.
- Helicopter tour of developments.
- Interview with case study (owners of one home on development).
- Interview with Russian estate agent working in Moscow and London (who was also a buyer on the development).
- Interview with Irish property consultant based temporarily in Moscow.
- Interview with English estate agent based permanently in Moscow.
- Tour of different types of development in the city.
- Photographs provided on site.

This itinerary allowed the journalist to have ample opportunity to get a thorough briefing in the local market, to have case studies to use as appropriate in a story, to meet the developers and other objective property experts who were familiar with territory that is highly unfamiliar to the journalist and his readers. Because of distance and language issues, the PR left much of the organisation to the agents and developers.

After the trip: The PR liaised with the journalist regarding the story, which was written and filed within 24 hours of the trip concluding. There was agreement from the PR that the story would not be given to any other national newspaper until after it appeared in *The Times*.

The result? Within two weeks there was a 1000-word article spread across two pages of *The Times*, including three photographs and contact details for the developer. Because the journalist involved was a freelance with good connections, he also got coverage in *Homes Overseas* magazine and *Estates Gazette*. So the PR, Jeremy Dodd of Barratt Howe public relations agency, secured extensive positive publicity for his client.

Another successful press trip was in January 2005 when a journalist was taken on a three-day tour of Edinburgh's and Glasgow's new developments by Knight Frank, which has acted as a consultant to many residential and mixed-use developers in city centre locations, and which has developed estate agency expertise in selling new-build city centre homes.

Preparation: Knight Frank and the journalist agreed on a visit to two cities where the firm's new city centre developments had been largely under-reported in the mainstream and industry press. The journalist was already confident that trade publications like *Estates Gazette* and *Show House* would carry pieces, and that broad information about the Scottish residential market would be useful for subsequent pieces in mainstream newspapers.

Trip details: Journalist met at Edinburgh airport by PR and Knight Frank's Scottish director to start an itinerary that included:

- Tour of current and prospective new-build residential sites in central Edinburgh.
- Tour of rural developments and conversions on the edge of the city.
- Meeting with niche local development company and tour of two of its sites.
- Trip to Glasgow and discussion with Knight Frank staff about city market.
- Tour of city centre developments and of key Glasgow suburbs.
- Tour of large development site at Glasgow Harbour and informal meetings with sales staff from a range of developers (not all represented by or contracted to Knight Frank).

In 48 hours the journalist had met almost all of the key new-build players in the two cities and had at least a cursory understanding of the local market, of important issues including the variations between England and Scottish planning systems and their different approaches to development, plus a look at important sites. Again, the Knight Frank Scottish team of agents and planners had a big hand in selecting who should be met and what should be seen.

After the trip: PR liaised closely with journalist to provide photographs as appropriate.

The result? As before, the journalist concerned was well-connected and within two months *Estates Gazette* carried a profile of one Edinburgh development company, *Show House* carried a critique of the Scottish new-build market, and the *Sunday Times* and *The Observer*'s residential property sections each included references to developments seen on the trip.

But failing to stick by the principles outlined in chapter 3 can lead to problems, as in the case of an ultimately abortive press trip arranged by a small and inexperienced PR agency in autumn 2005.

In this instance, a journalist was invited to see a new development on a Caribbean island by a PR, promising it would be exclusive. The journalist regularly wrote for a string of newspapers and approached *The Times'* Bricks & Mortar supplement to secure a commission — the PR did not know the journalist had selected this particular publication.

Then the problem occurred: two months earlier, the overseas editor of Bricks & Mortar had been directly approached by the same PR about the same story, again on the basis that it would be an exclusive. Of course, the newspaper and the journalist then each realised that the supposed "exclusive" had been touted to at least the two of them and — for all they knew — several other editors and journalists too.

The result? Both the journalist and the newspaper agreed to abandon the story, and the reputation of the PR was damaged as a result of what was interpreted as sharp practice. The trip, of course, never took place and what would have been a substantial two-page spread in the newspaper was never written.

So far, the press trips described have been for solo journalists but it is possible for more than one to be invited along, providing they are not given the same story to write for directly competing media.

So you could have four journalists, for example — perhaps one writing for a national newspaper, one for a trade publication, one for a lifestyle magazine and one for an online website. These journalists would have different needs, different stories and different deadlines, and the client would probably get at least four good stories.

If the four journalists were all national newspaper writers, the client may temporarily feel more important by getting such senior journalists together but, in reality, as soon as one got a story into print, the other three would back off. This is because, as we have seen, editors are covetous of stories and do not want to be seen as having a second and later stab at the same story.

Press trips and exclusives work, as with so much PR, on the basis of quality being more important than quantity.

> Editorial coverage is a means to an end, not an end in itself, so the amount of coverage achieved is less of an issue than whether it is appearing in the publications most likely to influence potential buyers

according to Tim Stanley, a senior PR.

> Short editorials with telephone numbers attached on the residential property pages of *The Mail, Express* or *Independent,* for example, can result in healthy levels of enquiries from readers who are would-be purchasers. Colourful spreads in all the monthly magazines can look impressive and may produce leads, but knowing which ones are of real commercial value takes experience

he says. "What will impress the sales director more — a thick sheaf of press cuttings or a neat bundle of sales enquiries?" asks Stanley.

There is one other issue to remember about press trips. This concerns follow-up.

Often a journalist will only be on a trip because he or she has a commission to write a story about the new development or new location irrespective of whether what he or she sees is good or bad.

It does happen from time to time (but less often than one might think, thank goodness) that the development or location visited is of poor quality. Some journalists will write about it and say exactly what they feel but others will be more discreet, and a good PR will respect that.

"As journalists, we need to know what's expected of us before we attend a press trip," explains the editor of one property magazine.

"If there's no prerequisite [to write a story] then there's nothing worse than being hassled or even bribed on your return. If a development or property is not up to scratch, then we simply won't write about it, no matter how much guilt we are made to feel," she says.

"PRs should realise that this is better for their clients, as rather than slating them we just omit them. It would be very helpful if PRs understood this editorial principle rather than assume coverage [will happen] just because we have attended something."

Don't panic

So far, we have looked at handling direct contact with journalists, including those where the interview or story appears to be hostile. If you have handled things well, and called in PR assistance, the story's impact may be at least minimised or possibly dropped completely. But what if it appears?

The key lesson is "don't panic".

Case study — Clucas Communications

Peter Clucas, MD of his property marketing company, says a publication once printed a front-page story about a client that suggested — incorrectly — wrong-doings within the firm.

The firm's board wanted to contact its customers and those partners with which it had professional relationships to allay their fears. But Clucas advised them to step back.

"I pointed out that many of the firm's clients would not have seen the article at all, and that many that had would see it just as a scurrilous headline designed to attract attention. To contact everyone would simply draw attention to the story," he says.

Because there was perhaps a case for libel or defamation, lawyers were contacted.

"The meeting with Counsel confirmed that, in his opinion, the journal may have a case to answer. It was made clear to us, however, that the headline alone was not enough. The whole article would be taken into consideration in any ensuing action. It is assumed that any reasonable person will read everything related to the story before reaching an opinion."

On Counsel's advice, Clucas contacted the publication's editor, whose response was measured and suggested that it, too, had sought legal advice.

"A meeting between my client and the publisher of the journal followed where an agreement was reached. A retraction was published in the form of a more detailed account of the events. This added missing facts and distanced my client from the story. A financial settlement was also agreed out of court. As far as we can determine, no long-term damage was done, either to my client's company or, perhaps more important for a PR, to our relationship with the journal. Respect was built between us that has, if anything, strengthened our relationship," says Clucas.

The lessons? He says the importance in this form of crisis management is to read the whole piece, not just the headline; to avoid drawing attention to a problem; to take legal advice early; to plan for and be prepared to agree a settlement; to remember that journalists take legal advice before they publish. One more thing — don't panic.

What's this got to do with property? By taking an unconventional approach to what would otherwise be a predictable "corporate" picture, Rok's chief executive and finance director anticipated they would receive more publicity. They were right — this shot was used in the *Financial Times*. *Source:* Rok property solutions

This is a good photograph, well taken and the property is interesting enough. But because it was similar to many other Spanish villas, it received little attention from newspapers and magazines when it was marketed in the UK. *Source*: Purple Cake Factory PR

... but when the media relations advisers working for the sales agent instead used a case study of a buyer — accompanied by shots of photogenic buyers — more editors became interested. *Source*: Purple Cake Factory PR

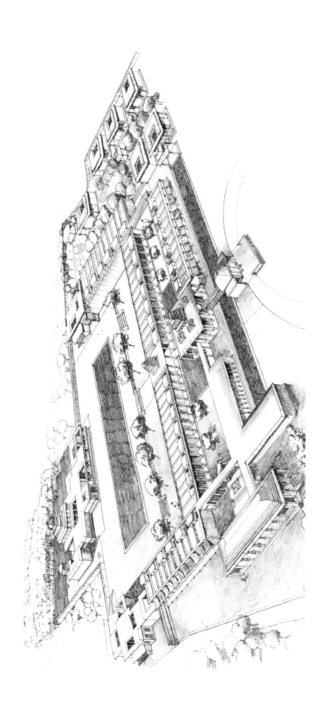

Media relations companies can be frustrated in their efforts to obtain publicity for a client's development if the client, in turn, does not pay for high-quality computer graphic images or computerised drawings. Without them, editors will not use a story about the development — which is what happened with this hotel development when it was first marketed. This drawing was not good enough for editors. *Source:* Nightingale PR

Some computer graphic images (CGIs) are very good and do attract publicity. This CGI of the Fabrika residential development by Circle 33 Lifespace was used in *The Times'* Bricks and Mortar property supplement. *Source*: Barrett Howe PR

When you see the real, completed Fabrika property you can see how sometimes a good CGI is almost as accurate and impressive as a picture of the finished product. *Source*: Barrett Howe PR

Developers working in emerging markets usually want to keep overheads to a minimum so tend to produce low-quality computer graphic images (CGIs). This CGI of a Bulgarian resort may have been sufficient to put on a website for investment buyers but no newspaper used it during a marketing campaign in the UK. *Source: TRMA*

Interior shots are perfect for specialist articles about design but most magazines and newspapers want external shots that give a feeling of size and location. Most interior shots could be any sort of apartment in any location. Where is this? What sort of apartment block is it? The same applies to internal office shots, where commercial property publications really want external pictures. *Source:* City Lofts

Identifying an angle — the media relations firm helping to promote this mixed-use tower in Manchester realised that many articles in the specialist and mainstream look at environmental aspects of development. This is not necessarily the most important aspect of the property to the developer, but it would never have made it into publications had it not been for photographs of solar panels attached to specialist press releases emphasising the "energy" angle. *Source*: Taylor Woodrow

Even if the property is not glamorous, unusual approaches can make a picture interesting. This is a "£60,000 house" demonstrated at a conference in 2005. It was designed in 14 weeks and built in 36 hours. English Partnerships, the quango behind the creation, widely circulated this interesting "action" shot of the property *in situ* at the event. As a result, it won substantial publicity for its scheme to build low-cost homes. *Source*: English Partnerships

Proactive PR — Promoting your Key Messages

The previous chapter focused on how to respond most effectively to requests from and contact with journalists. In this chapter, we consider the various tools and techniques used by media relations professionals to take a more proactive approach — to prompt those calls and interactions with journalists and to generate positive coverage in situations where there is not any "hard" news.

Raising awareness of the need for media relations

The starting point to taking a more proactive approach to media relations (or even being more effective at reactive media relations) is to raise the level of awareness, understanding and co-operation from the people in your firm from whom you need information, time and co-operation. This is because even the most proactive PR professional will fail miserably without a significant number of senior people behind him/her promoting and supporting the proactive media relations effort.

But before we can engage the Board or other members of the firm in a discussion about media relations and their contribution to generating positive media coverage, we need to get their attention. If we can show them some clear pain or a desirable potential gain, then we will do this quickly and effectively.

Here are some thoughts on how to get the need for proactive media relations onto their agendas.

Competitor watch — Scan the key media of interest on a regular basis. Cut out or note down the coverage obtained by two or three of your major competitors. Report regularly on what competitors have said and how much coverage they generated. It should not take long before people get annoyed with reading what their competitors are saying and doing, and there will be plenty of comments such as "But we do that better ..." or "But they're talking rubbish ..." or "We have much greater knowledge in that area ...".

Champion — Identify an individual within your organisation who is keen to raise his or her media profile, who has the right attitude and something interesting to say (see chapter 3 for guidance in this area). Devote as much time as you can to working with this person, understanding what they have to say, who would be interested and in how the messages can be conveyed. When others see the increased profile for one individual, they will be keen to see the situation replicated for others — either for the corporate good or for old-fashioned professional rivalry.

Client/occupier research — Identify a representative sample of your clients and/or existing or potential occupiers. Formulate a structured research exercise to obtain their views and perceptions on how your company and its products, services, investments or buildings are perceived. Ensure you ask questions about what media they read, hear and watch and what they picked up from recent coverage. There is nothing like hearing about your lack of media profile from the horse's mouth to raise interest in taking a more proactive approach to media relations.

Media/journalist research — Similarly, identify some key journalists from a selection of your target media. Contact them (if you have some information for them to use and/or some ideas for possible articles so much the better) and ask them a series of short questions about their views of your organisation. Ask them to compare their knowledge and views of your organisation with that of specific competitors in the same marketplace. You might find it easier to commission an external media relations agency to undertake this work for you as a project, particularly if you do not have existing relationships with the journalists in question. The objectivity of the external agency will add credibility to the results and they will be able to include a series of recommendations for action as part of the study.

Education and training

So you have put media relations into everyone's minds. The task now is to educate them on what can and can not be achieved. Manage expectations from the start to avoid angry and disappointed accusations later on. Help them understand what they must do to support the activity.

Again, there are a number of different ways that you can implement a large or small and subtle internal campaign to get people engaged in the media relations process. Here are some ideas.

Guest speaker — Arrange for a key journalist or even someone with a high media profile to attend one of your major meetings as a guest speaker to give their views on your firm's profile and approach to media relations or to talk about your competitors who are good at proactive media relations.

Departmental/team meetings — Request a slot at the next departmental and team meetings and prepare a short introduction to media relations to present to meeting delegates. Once you have done your presentation, lead a discussion to generate ideas for articles and news stories. Try to put media relations onto the agenda so that it is discussed at all subsequent meetings. Ideally, you should take along to each meeting a file of the cuttings generated by the department — and possibly with comparison material from your competitors.

Internal training sessions — Invite people to attend a lunchtime or early evening training session on media relations. Provide sandwiches and drinks as an incentive. You might also invite along a senior person internally, or a client or journalist, to add another perspective.

Bring some solid examples of positive media coverage obtained by your firm or its competitors. While you will need some formal material to present to them, you should also identify some issues that can be discussed and debated as a forum. For example, break into groups so that people can think about what key messages they would like conveyed to the media, or to identify interesting developments within the firm that might usefully be communicated more effectively. Ask them to identify which media and magazines are most influential in their markets — there may be considerable difference between those selling beautiful new homes to wealthy people, those promoting cut-price sheds to small industrial companies and those selling premium retail space to major brand owners.

Take along some examples of the media you target so they can develop an understanding of the sorts of news and feature material required. You might add a competitive element and get the various groups to see who can generate the most ideas for articles or who can produce the most synopses for potential articles. Ensure they have some brief notes to take away with them and circulate a summary of the key ideas and issues discussed at the session afterwards. If you can schedule some of the identified actions into a campaign plan that can be monitored, then even better!

External marketing/PR course or event — Take some of your people along to marketing and PR events where media relations is being discussed.

PROFILE, a networking group for people involved in property marketing, has an excellent programme of events and frequently invites members of the national, property and other media to provide talks on how to maximise the possibilities for coverage. Profile events have the added advantage of being attended solely by other property people — both the property professionals and the marketing/PR professionals.

There are other groups where professional staff and marketers get together — for example, the Professional Services Marketing Group and the Marketing Partners Forum. There is also a wide range of commercial training courses to help people understand the techniques and approaches to media relations. These are available from training organisations that specialise in marketing and PR and also those who specialise in the property sector.

You can either pay to send some of your staff to these events — where they have the advantage of meeting and networking with people from other businesses and industries — or arrange for an in-house presentation of the course so that a number of people can attend.

Best practice and "how to ..." guides — Within larger organisations in particular, it can be useful to document guidance on how to tackle common media situations in simple terms that are directly relevant to your organisation. These guides can then be sent to people when they have a particular media relations activity to tackle or made available on an intranet as a reference source. In addition to saving the internal or external media relations professional time explaining the basics in each situation, it will also ensure that there is some consistency in approach in how media relations is tackled. These guides are also an excellent way to convey the firm's policies and procedures (see below) in the media relations arena.

Regular internal communications — Having got everyone's attention on media relations and generated a good level of understanding about what is involved, you then need to maintain the motivation and momentum. This requires a concerted and sustained effort in internal communications, which is dealt with in the next section.

Intensive "front of camera" training

Having raised awareness and developed understanding among a broad range of people within your organisation, you now need to develop some real expertise and experience in journalist-facing situations. This means you must identify those individuals within the organisation who are most likely to be in "on the frontline" and in "front of camera" situations. You then need to arrange for these people to undertake intensive media training where they can practice presenting their story, answering questions and being interviewed by journalists. The best way to obtain such training is from experienced specialists who can field real journalists to simulate the interviews and provide actual studio conditions (both radio and television), so that the experience is as real as possible.

These courses often provide facilities to video the exercises and role plays undertaken by delegates so that they can observe their performance and the improvements that have been made or that need to be worked on.

Internal communications — ensuring a flow of information and opportunities

Once there is a reasonable level of awareness and enthusiasm about media relations within your firm, you need to ensure that the relevant people take the time to advise the media relations professionals (both internally and externally) about topics and changes that might be used in the media relations programme. This is much easier said than done as your property professionals will be focused on doing their day job of building projects, selling and negotiating deals, undertaking work for clients, so media relations will rarely be uppermost in their minds.

As such, the media relations professional needs an ongoing internal communications campaign to keep talking to all those who are likely to have the news and material that can be used for generating publicity, and to remind everyone about the sorts of information that is needed to maintain a good media profile. Similarly, it is important for people in your firm to know at the earliest opportunity what information is being sent to the media and what coverage results so that they are not caught unawares when one of their clients or contacts spots the coverage and asks them about it.

An internal communications campaign is similar to an external campaign. Within your firm, there will be different people with different levels of awareness and involvement in media relations and with different contributions to make. So you need to "segment" your internal audience as much as your external audience and tailor the messages and information to their particular needs.

Here are some of the techniques that help keep media relations uppermost in the minds of those people on whom media relations professionals are reliant for information and time.

E-mails — Regular, short e-mails about coverage achieved by the firm or its competitors will act as a reminder of the sorts of information that is needed by the media relations professionals and as a demonstration of the sorts of benefits that can be achieved when people take the time and effort to obtain the information needed (and the necessary permissions from their clients or contacts to use it). You should also make sure that all press releases are circulated to everyone internally at the same time as they are released to the media. E-mail is also a fantastic tool to issue requests and alerts to people about forthcoming features or issues on which you have received requests from journalists. Some media monitoring services provide e-alert systems that allow people to select the companies on which they would like to receive updates on new coverage — setting these up for people to ensure they see the results of the firm's media relations efforts keeps it top of mind. E-alerts notifying people of competitors' or clients' coverage can be just as effective.

Press clips — While being mindful of copyright laws (you may need to pay a fee to the Newspaper Licensing Authority) it is a good idea to keep a file of all the press clips that mention your firm. You often see such folders in the reception areas and meeting rooms of organisations where they act as a good source of information about the firm and its activities. Some firms complete weekly, monthly or quarterly compilations and have these more widely available on staff notice boards and in staff coffee or dining areas. Some firms have areas of their intranets or websites dedicated to keeping electronic or soft copies of their clippings. Many intranets have a section on their home page showing the organisation's latest media coverage. Again, please be aware of the relevant copyright rules.

Intranet — Many intranets have the ability to flash news items on the home page and drive users to sections of the intranet where press releases, press coverage etc are stored. Use these devices to keep people aware of what information is being sent out to the media and also what coverage results.

Website — If those within your firm use the firm's website a lot, ensure that press releases and media coverage are posted onto the news areas of the website on a regular basis. Ideally, there should be an area on your home page where news is shown. Many websites have a specific media or press centre where such information is kept. As well as being a useful source for your staff and the media, many clients find it helpful to check up on the news section of their landlord or adviser's websites.

Team meetings — Ensure that every team meeting has a slot for marketing and business development issues and that part of the time is allocated to discussing these issues relating to media relations opportunities. Media relations professionals can deepen their knowledge, strengthen internal relationships and spot opportunities for media coverage by regularly attending the team meetings of different departments within the firm.

Reprints — Media relations coverage can often be recycled and used as promotional tools to supplement the brochures and other publications that your firm produces. Third-party endorsement is always more credible than what you write in your own promotional materials. This is particularly useful where people within your firm have written or made a substantial contribution to an article. The publisher should be contacted as there is usually a requirement for you to obtain permission — and sometimes to pay a nominal fee — before reproducing articles to give away or mail to clients, customers and contacts. The sight of reprints in information packs and in reception and meeting areas acts as a constant reminder to people of the value of generating good media coverage.

Progress reports — Either as part of your regular management and marketing reporting process or as a separate media relations exercise, you should prepare a regular report on what media relations activity has taken place, what coverage has resulted and how this compares to your performance in the previous period and against what you had planned to occur. This topic is covered more extensively in chapter 6 on measuring effectiveness and results.

Media relations meetings — The final, and most obvious, way to ensure that the need for media relations is to maintain and obtain a flow of information is to have regular or ad-hoc meetings with the sole purpose of discussing past media relations activities and planning a forward programme of media relations activity that will integrate with and support other promotional activities. Such meetings should have a clear agenda, be kept as short as possible and result in concise minutes that list what actions have been agreed and by whom and on what timescale. A central role of many media relations professionals is checking that people complete the actions they have suggested.

Develop a media relations plan

Hopefully, there will be business and marketing plans for your firm and/or its various departments or developments that you can use as a starting point to plan your media relations activities. If not, then you will have to do more work in order to prepare your media relations plan.

Depending on the nature of your organisation and/or the developments you are promoting, there may be a need for separate media relations plans addressing the "corporate" and the individual markets, services, developments or properties (see chapter 1 for more information on the differences between corporate and product PR).

Plans do not need to be long or elaborate. A simple structure and a summary of the key thoughts will suffice, although there must be a reasonable degree of detail in the action plans (and budgets). It is also worth remembering that the process of developing a media relations plan is often more valuable than the final plan that is produced — a clear case of the journey being more important than the destination.

Some people argue that media relations is purely opportunistic and dependent on changes in the market and deals and that, therefore, a plan is inappropriate. We agree that there is a strong element of opportunism in media relations and that you must remain alert and flexible in order to respond fast to opportunities as they emerge. However, the media relations plan is important in analysing the situation, setting some clear goals and identifying what actions should be taken in addition to any opportunities that are seized when they arise. They also play an important role in communicating to others in the firm what will be happening and why, and what the media relations is trying to achieve.

Media relations plans come in many shapes and sizes. Those produced by media relations professionals will undoubtedly be far more structured and detailed than those produced by property professionals. However, as a guide, below are the sorts of headings that you should expect to find in a media relations plan.

Media relations plan outline

Executive summary
Where are we now? — Analysis of the current situation

- Past/current media profile.
- Competitors' media profile.
- Perception of clients, contacts, influencers, referrers.
- Business/marketing plan highlights.
- Summarise the strengths, weaknesses, opportunities and threats.

Where do we want to be? — Setting some goals

- Links to business and marketing objectives (from other plans).
- What do we hope to achieve with media relations (short, medium and long term)?
- Specific objectives against which we will measure progress and results?
- What are the key messages we wish to convey?
- Which media (journalists) are we targeting?
- The systems required to monitor and measure progress against the goals.

How will we get there? — Agreeing a strategy and action plan

- Preparatory work (information needed, support materials required, training etc).
- Resources to be used (external PR agents, internal PROs).
- Integration with other marketing and business development activities.
- Key spokespeople.
- Potential stories — press releases and articles.
- Meetings and interviews.
- Other media relations techniques and approaches.
- Planned campaigns.
- Management, monitoring and measurement systems.

Action plan (timescales and responsibilities)
Budgets
Appendices (supporting information)

In the strategy section of the plan, you will need to decide what type of broad approach you will be adopting for your media relations. For example, some people adopt a regular "drip drip" approach, ensuring that there is a regular flow of information going out to the media while others adopt a "burst" approach, where there is a major effort to generate a lot of coverage in one go across a range of media. A combination approach is the "noise and peak" approach — regular information to maintain a minimum level of "noise" about your firm plus occasional bursts of media activity within campaigns to provide "peaks" of awareness.

Figure 5.1 Noise and peak approach to media coverage

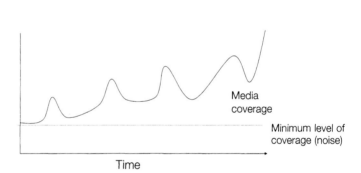

For those who are really doubtful about the value of producing and using plans, you may wish to adopt a more action-oriented approach and produce a simple project plan of the sorts of media relations activity that you plan to undertake. Figure 5.2 illustrates this approach.

Formulate policies and procedures

Having set out what you hope to achieve in your media relations plan, you ought to anticipate some of the potential problems that might occur to avoid both crises and/or repetitive situations. There are a number of areas where you may wish to discuss, agree and then communicate the firm's overall approach to media relations. For example:

Figure 5.2 Example media relations project plan

Activity	J	F	M	A	M	J	J	A	S	O	N	D
Press release	✔		✔	✔	✔	✔			✔	✔	✔	
Article – Magazine 1		✔			✔		✔				✔	
– Magazine 2			✔		✔		✔		✔		✔	
– Magazine 3		✔		✔		✔			✔		✔	✔
– Other			✔			✔			✔			✔
Journalist meeting	✔				✔				✔			
Media tour of site							✔					
Photo/celebrity event							✔					
Journalist round table									✔			
Other			✔			✔						✔

Spokespeople — Some firms have policies that indicate which individuals may have contact with journalists. These may be the most senior members of the firm (eg, board directors), media relations staff or other people who have been through a media relations training programme. While these policies are effective at ensuring that only those with the relevant knowledge, authority and skills work with the media, it also enables an organisation to control, co-ordinate and monitor contact with journalists more easily. In many cases, the policies also act to protect junior and less experienced staff from being in situations with journalists that they may find uncomfortable or that may jeopardise their positions.

Disclosure (financial) — The Stock Exchange rules governing the disclosure of financial information will be familiar to the specialist financial PR people who work with publicly listed companies. For private companies and partnerships, they must decide if and how they will release financial information to the media. Some partnerships disclose income but not profits, some private companies release summary information while others provide detailed statements. The critical issue is to decide what your position will be in advance and to ensure that it is adhered to.

Disclosure (clients) — You may need procedures to control the release of other — potentially sensitive — information to the media. For example, you should ensure that you have permission from third parties, such as joint venture or consortium members, before releasing information about strategies and plans. Similarly, you should have clear procedures for obtaining permission from clients or tenants before releasing information about deals and transactions.

Sign-offs and approvals — You may have procedures for who is authorised to sign-off press releases or media statements and impress upon them the need for proof-checking or verifying the broader implications of releasing any information. As well as providing a mechanism to control the quality of the information being released, this also ensures that another pair of eyes — typically those without direct involvement in the matter — reviews the information and identifies any gaps or weaknesses. Some organisations have policies that require at least one board member to sign off all information that is released. You might also ensure that all press releases are circulated to key members of the firm before being released to the media — back to that internal communications point again.

Legals — When providing information to the media, you must be aware of a host of potential legal restrictions. As mentioned above, financial information will be subject to Stock Exchange rules. Obviously, confidentiality and copyright must be considered. The contracts governing particular deals may also be relevant. If conveying personal or negative information, you must be alert to issues such as privacy, human rights and defamation (see p139). Understand where legal issues may be a problem and ensure that the relevant policies, procedures and access to expert advice are in place.

Crises — There will be occasions when there are emergencies, sensitive situations that are of major interest to the media and potentially damaging to an organisation (see chapter 3). In times of crises, it helps if there are clear procedures about who will do what and when with regards to the media. In emergencies, it is helpful if a single senior person is available to speak to the media — whether this is on television or radio (which may require them to go to the location of the problem) or on the phone. You might also use your external media relations consultants to provide a focal point for taking all enquiries and allocating responsibility for action. In times of crises, the need for fast and good internal communications is vital, so there should be procedures in place for this. You might also want to include the need to prepare and distribute internal briefing documents or standard sets of questions and answers so that everyone remains aware of what is happening and the media receives a standard, accurate response to their questions regardless of who they deal with.

While these and other procedures may appear bureaucratic and to cause unnecessary delay, they can actually speed things up and provide some control over what information is released and when. Identifying key individuals with responsibility for media relations can often increase the speed with which an organisation responds to media enquiries and opportunities. Checking and sign-off procedures can help avoid negative or inappropriate coverage and protect junior staff from embarrassing mistakes in print.

Case study — Media relations and the law

Simon Smith, one of five partners in leading law firm Schillings who all specialise in this field, spends the majority of his working life — and a significant amount of his weekends too — dealing with high-profile media crises. This year, he has won libel actions for Cameron Diaz, Justin Timberlake, Rio Ferdinand, Kofi Annan's son, among many others, and he has just commenced similar actions for actresses Teri Hatcher and Kate Hudson.

His firm acts in media management for the likes of Naomi Campbell, establishing in a case brought for her against MGN the UK's Privacy Right in the House of Lords, and won at trial this year in a libel action for Roman Polanski in which he gave his evidence to the jury by video-link, another legal first. The firm does not just represent celebrities, as the same skills are deployed to protect companies and their officers and employees. For instance, in the past two years Simon has acted for major organisations in media matters such as The London Stock Exchange, Royal Mail, Post Office, Rover and Grant Thornton.

In property-related media disputes during the same period, Simon has acted for the architects Foster and Partners, influential property investment funds, Spanish real estate agents Ocean Estates, and various London-based developers.

"Good PR policies (such as pre-crisis planning) and practice should ensure that media crises requiring legal remedies are infrequent. However, there is always an outside chance a media crisis could happen and it is better to be prepared on what legal remedies might be available. For example, a leading property figure, who usually shuns publicity, was splashed all over the tabloids as he initiated his divorce.

There are four areas of law that might be applicable in a media crisis:

- Defamation
  Libel is where inaccurate or untrue information has been printed or

broadcast (including on the internet). The legal test requires that the material is published and the object (a person, such as a director or an organisation, such as a company) is identified and that it reduces reputation. Both human feelings might be affected and damages may be due, such as for lost commercial contracts.

Editors — and at Schillings we have dealt with them all on numerous occasions — will look at any story to see a) what supporting evidence they have and b) the likelihood of the target suing them. Having expert legal advisers to take control of the situation immediately is an enormous help. If it is too late to prevent publication of a negative story, then there are numerous remedies, such as prominent retractions and apologies or even financial settlements, that might be possible.

- Privacy
  This concerns the misuse of private information that is then published or broadcast. Typical situations are when ex-employees disclose confidential information, such as trade secrets or there are leaks. This can also be applied in situations to prevent people using material that is filmed on private property — for example, in shops or on private land. It is an effective tool to prevent intrusive stories about members of the board, or even their wives/husbands and children.

- Copyright
  Whether material is handwritten, dictated onto a recording device, video footage, an e-mail or a photograph, the copyright remains with the originator. Therefore, it is unlawful for people to publish such information without the prior permission of the copyright owner. Copyright law can sometimes be used in media crisis situations to prevent the publication of letters, notes etc around which a story is based — thus reducing its attractiveness to the media. Public interest is not a defence against breach of copyright and many journalists are unaware of the legal situation in such cases.

  Where there are fears of documents being leaked, there are numerous techniques that can be used to protect documents or, at the very least, ensure that you are aware of the precise source.

- Trademarks
  There are situations where you or your organisation's name or image might be used inappropriately to infer an association or endorsement in situations where you have not given permission. Trademark law can be used to prevent this, or alternatively common law remedies such as "passing off" apply.

Schillings' *Protecting your brand* — *Preparing for crisis* document offers the following advice:

Always:

- Find out what a journalist's deadline is — how much time have you got to respond?

- Take their name, mobile number and their newspaper/magazine.
- Fill out your crisis directory.
- Call them back.

Never:

- Say "no comment" — ever read an article that makes 100 accusations and the accused can only muster a feeble "no comment"?
- Assume that an "off the record" comment will not come back to bite you: you will read your comment as a quote from "an insider" or "someone close to the company".
- Be afraid to politely put the phone down. Journalists know how to make people talk. The longer they keep you on the phone the more likely you'll say more than you intended.
- Lie. If they catch you out, they will never let you forget it.

Prepare press releases

Chapter 3 considers "What makes a good story" and contains examples of good and bad press releases. This section explains the process for preparing a press release — which is one method of getting a good story out to the media.

It is worth considering for a moment the purpose of a press release. Some people seem to think that the media will simply reprint the press release as it appears — but this is rare. The purpose of the release is to provide information to journalists in a form that they find easy to use and to prompt them to contact the relevant people in your organisation to obtain additional information that will enable them to pursue the "angle" that is of most interest to them and their readers.

Press releases must be written in a particular way. It is difficult for people who are expert at writing other types of document — whether technical papers for clients or promotional material for other marketing activities — to adapt to the style of writing press releases.

Press releases must contain the following information.

- *Stationery* — Most organisations issue press releases on letterhead paper — providing contact details and also legally required information, such as lists of directors or partners, the registered business address and any regulatory information. Many organisations have special press release paper printed so that it gains attention among all the other pieces of papers and is immediately apparent that it is information for the media.

- *Layout* — The way to set out a press release is very specific — although the reasons behind the format are historic — from the age before computers were invented. These are fondly known by ancient journalists as the days of "hot presses". You should use 1.5 or double spacing — this is because originally journalists made editing marks directly onto the release and then passed them to the typesetters to set. This is also the reason that press releases should only be typed on one side and at the end of each page — if there is more — you should type [MORE] and at the end of the release you should type [ENDS].

- *Date* — The date the information is released must be shown. Naturally, journalists will only be interested in information that is timely, so ensure your release is issued on the same day as the date shown. Be aware of time differences if you are dealing with international media. Take great care not to embargo information (ie, release information to the media before you wish for it to be published) unnecessarily as the advent of the internet means journalists may inadvertently or deliberately release information before you would wish.

- *Title* — Avoid the temptation to write a headline for the recipients — they are the journalists. Furthermore, the sub-editor will write the headline when the piece goes to print — not the journalist writing the actual story. The title should be short, in an active tense and clearly describe the content of the release.

- *Language/style* — Avoid superfluous "puff" in press releases — journalists do not respond well to marketing speak. Keep sentences short. Keep the information as factual as possible. If you make claims, ensure they are substantiated with the relevant facts and figures and sources. Avoid the use of capitals for job titles. One to nine should be written as words, 10 and above as numbers. Organisations should be referred to in the singular. You would do well to invest in a style book — *The Economist style guide* is recommended.

- *First paragraph* — The first paragraph must be a summary of the entire story and answer all the questions — who, what, where, why and how — quickly. Journalists will spend only a few seconds reading the title and the first few lines of a release — if you are lucky — so do not hold back key information until the end as they will rarely read that far.

- *Subsequent paragraphs* — Each subsequent paragraph should provide more information on each of the points covered in the first paragraph — starting with the most important information first. The least important information should be towards the end. This structure is for historical reasons. When originally setting a piece, the story would be edited to fit the space available and paragraphs towards the end would be omitted first if space was tight. A good test of your release is to check that it stands up as only a first paragraph, then as the first two paragraphs, then as the first three paragraphs and so on.

- *Quotes* — Every release should contain at least one quote from an identified individual. Ensure that the name and position of the person is clear at the start of the quote. And that the quote is short and interesting. Provide contact details for all those people who are quoted in case the journalist wishes to talk further to the individual.

- *Contact details* — There should be at least two contact names at the end of the release. The contact names should show the name, position and organisation of each person and provide a direct dial telephone number and also an e-mail. If it is a major or international story where out of hours contact is likely to follow, provide mobile numbers too.

- *Editor's notes* — Supporting information that might be helpful background to a journalist who is not entirely familiar with the subject matter or your organisation should be provided in this section rather than the body of the main release. Typically, you might have a short description of your organisation, background or technical information relating to the item in the release, biographies of those who are quoted or who provide their opinions.

- *Photographs* — The need for good photographs (and short captions with a date) is addressed in chapter 3. You will increase the chances of your release being used significantly if it has an interesting photograph.

While a press release contains general information, it is a good idea to prepare different versions if you are sending it to various types of media. For example, you might include a regional angle for the local consumer press, a business angle for the business trade and technicals and a financial angle for the financial press.

Very often — for example, when reporting major transactions — there will be many parties involved who will all wish for their role, contribution or views to be included in the press release. For example, in commercial property transactions this might include the developer, a consortium of contractors, the funders, the major tenants, the surveyors/agents and even the lawyers and accountants. In these situations, it is helpful if one party (perhaps the media relations agency for the project or development) co-ordinates the preparation of a release and works with all third parties to obtain their input, convey accurate details about their role and contact details, and approve what information is released to the media and the timescales and other logistics. In such situations, it is also advisable to agree who will act as the major spokespeople on different aspects of the project.

You should also consider preparing other background briefing papers for journalists if it is a major or complicated story. These briefing papers are not intended to be used in coverage (although you should be aware that they might be) but to help the journalists understand the context, background or technical details. The same care in drafting and approving these materials is required as with any other material you release to the media.

For a particularly important or major story you will also need to consider a number of other activities in addition to preparing and distributing a press release. For example, it is good practice to prepare — for internal use only — a Q&A document that lists all the possible questions (both positive and negative) that the media may ask as a result of the release. There is good value in helping everyone see the story from the media's point of view, anticipating all the possible questions that might be asked and providing some guidance on what responses might be given. In addition to preparing all members of your organisation for fielding a wide variety of questions — which will boost their confidence — it will also ensure that all media will receive a standard and consistent story on all the critical points.

Distribute press releases

While many people focus on the weekly property magazines, there is a wealth of other media that may be interested in your release. An overview of the media and examples are shown in chapter 1.

However, you need to avoid sending your release out to media that clearly have little or no interest. Therefore, you should develop a series of press release distribution lists that are likely to be interested in the different types of story that you issue. You can build these lists in a number of ways. You might purchase a hard copy media directory and extract the names and contact details of the relevant media manually. This is a sensible approach for a small organisation or one that does not undertake much media relations activity. Details of the various media directories are shown in chapter 7 (Useful information sources).

There are also a number of online directory services that enable you to search media lists and build your lists online. Some of these distribution services will also manage your lists for you and organise the distribution of your releases for you. Make sure you get to know the media that you use most often — and ensure that the release is sent to the relevant reporter — there will be different people handling news stories, transaction and deal information, regional and local information and financial information, for example.

Once the release is sent out, it is a good idea to telephone the main media to draw their attention to the release and to explore whether there are particular angles or additional information that interests them. However, if you simply phone up and say "Did you get my press release?" you are likely to receive an irritated response — many editors receive more than 300 releases a day, so you are demonstrating your lack of empathy (and arrogance) if you ask such questions.

The best way to promote interest in your release is by having good relationships with your target journalists. This means that you will know what will be of specific interest to each journalist, the time of the day or week when they are most receptive and how they prefer to receive and develop information. Advising journalists in advance of a major story and letting them know the release will be with them is a good idea — but reserve this for important information and beware of inadvertent exclusives (see chapter 3).

Organise press launches or conferences

Getting a bunch of journalists into a room together at one time is a prospect that makes good sense to a busy PR person. It means that they can save the time of their senior people who need only attend one event, it means that everyone gets the information at the same time and you are able to speak to many journalists in one go — very efficient, at least on paper.

However, this is fantasy land and most experienced PRs will advise against formal gatherings of journalists. The reasons are numerous, and have been discussed earlier. First, journalists are busy people and some are unlikely to turn up to an event even if they have said they will. Second, many journalists now want exclusives and that is difficult if everyone hears the same story at the same time. Third, how are you going to control things if one journalist decides to ask difficult or negative questions? Fourth, what happens if one of your key speakers screws up and all those present witness such embarrassment? Fifth, you really need at least one senior and experienced person available to speak to each journalist that attends — this will enable them to get from the horse's mouth the specific information that they need. Sixth, the few journalists with whom you have good relationships will resent the fact that they are being treated the same as the many other journalists who are present.

If you are determined to press ahead with a press launch or press conference, then there are some guidelines you need to follow.

Make sure you speak to the relevant journalists to explain why it is in their interest to attend. This is difficult as you have to reveal something about the nature of the story or information being released in advance and there is a danger it will be leaked. However, if you are releasing information about a merger or a significant transaction, they will understand the need for a meeting to obtain all the relevant information and gain access to all the key players at one time.

More common than press launches, at present, are conferences that are provided for mixed audiences — clients, potential clients and influencers — as well as the media. Having a good line up of speakers, who are recognised as experts as well as having research to release, will increase the attractiveness of such an event. However, if you are inviting the media along to a conference make sure that the speakers and the delegates are aware of this to avoid embarrassing misunderstandings.

Ensure that the speakers at the event are extremely well prepared for unusual or negative questions or questions about unrelated matters happening elsewhere in your organisation (or even in its ancient history) as well as the main story. Make sure the speakers are trained and experienced in dealing with the media, whether addressing them as a group or in individual discussions afterwards.

Prepare a comprehensive press pack. This should contain a press release with all the key material being conveyed as well as supporting and background information and photos and biographies of the speakers. Everything in a pack should be dated and contact details should appear regularly. You may find that you distribute more packs after the event, to those who did not turn up, than at the actual event.

Make sure you have as many minders present as well and that each journalist who attends is welcomed and spoken to. There is nothing worse than inviting someone to an event and then not making an effort to speak to them when they attend.

Use research

There was once a piece of research that showed that the lion's share of PR-generated press coverage was based on research. The property industry invests a huge amount in research — it is certainly a major element of the major surveyor firms' investment — yet is notoriously bad at using the research to prompt dialogue with journalists and press coverage. A critical issue appears to be a lack of awareness of how to present the key findings of the research to the media (as opposed to the main target audience of the research, which may be developers, business tenants or private residents).

The best way to ensure that your research is suitable for media coverage is to allow the media relations professionals to be involved in the design of the research project at the outset. This way, the information and desired results that will be of interest to the media can be designed into the research project.

If it is a substantial or technical piece of research, it may be necessary to prepare a summary of the key findings and include these in a covering press release. If possible, you should make the key tables available in PDF format so that they can be easily reproduced in the electronic or print media. Ensure that any tables like this have your organisation marked clearly as the source.

You can increase the value and credibility of your research by involving other organisations — for example, household known brands or leading academic experts. This adds credibility and objectivity to the research and extra interest to the media who you wish to write about it.

One of the most successful research approaches involves the creation and maintenance of a regular index — this provides a mechanism for your organisation to receive regular coverage on the results as well as an opportunity to provide expert opinions on the implications of the results.

Although you should summarise the main findings in a press release, you should always provide a copy of the original research report with the release — or indicate from where it may be downloaded on the internet. If the research report is available to the readers of the medium — either for free or for a cover price — ensure the release contains details of where readers can obtain it from.

A word of warning though — if your research story gets picked up by a national or a major publication, be prepared for a lot of demand for the research report. It is embarrassing for a newspaper or magazine to mention the availability of a report and then have readers complaining that they are unable to obtain a copy.

To illustrate the combination of conference and research, the following piece appeared in *Estates Gazette* in October 2005:

Headline: London rents set for 16% annual rise

Rental growth in West End and City offices could hit 16% pa within the next four years, writes Dalmar Farah-Niedenthal. According to speakers at the ING Real Estate annual seminar on corporate property, next year is expected to be a year of changing climate, with better prospects for investors on the horizon.

Speaking at the event at the Magic Circle in Euston, NW1, Ian Whittock — ING managing director of research and forecasting said: "In most markets 2006 will be the last year for occupiers to exploit their negotiating position. In the City office market it will be the last time the tenant holds all the cards.

"Development sites have been changing hands for £300 per sq ft, while new completions in 2006 will be the lowest for 20 years," he added.

Delegates heard that ING Real Estate was forecasting 3% rental growth for grade A offices next year, with overcapacity falling in Central London.

Head of corporate and retail marketing Christopher Tabor said rent-free packages across the office market would be significantly reduced in 2006, as corporate requirements grew.

"In the City, there are only seven buildings available over 100,000 sq ft providing grade A accommodations. Even Swiss Re is getting tenants. There is only 80,000 sq ft vacant, as opposed to 200,000 sq ft in December 2004," he pointed out.

Forecast out-of-town office rental growth for 2006 was below 2%, with rental values in the regional market expected to remain static.

Tabor went on to say that although rent-free packages were also being reduced in the out-of-town office market, the sector continued to suffer from overdevelopment of the technology boom.

In the retail sector, Monsoon and Next were said to be among 11 major retailers actively expanding while five anchor retailers were disposing.

However, Whittock added that intense competition would encourage many retailers to continue rationalising their portfolios.

In industrial and warehousing, a slowing housing market is expected to hit demand for bulky goods, leaving rental values to rise slowly on smaller units but remain static on distribution units.

Whether a company is small or large, relatively unknown or a respected institution, original and authoritative research can be a valuable peg for stories. In the residential world, Knight Frank capitalised when a senior research figure who had generated considerable national newspaper publicity for Savills, left to join a lower-profile company in October 2005. Knight Frank stepped up the PR focus on its own research head and stole the march on its rival for several months.

Case study — *Area Sq research*

The view of Guenaelle Watson, PR manager of office interiors company Area Sq, is that journalists are more likely to pick up a story that does not sound like advertising and which gives the company credibility, even if the subject matter may be controversial. Research is a vital part of that.

She says: "Area Sq's PR approach lies in a combination of building relationships with key journalists and creating PR opportunities. For example, in April 2005, we carried out a survey amongst 500 office owners and occupiers throughout the UK to evaluate what they thought of their current offices, and how future working practices would impact on their office space requirements and profitability. A summary of the survey results was sent out to the press and I received positive response from various publications, including *The Times*, who requested to receive the full survey results and printed the story. The Office Owner and Occupier survey has been a positive PR platform and Area Sq will definitely reiterate this initiative next year."

Case study — *RICS*

"You need something with 'legs' to get the best return on all the work that goes on at the outset," explains Andrew Smith, RICS external communications manager.

"Taking a position on government policy is a popular choice as government consultation and policy making often goes on over quite long time scales. Then, it's about establishing your organisation/firm as a recognised voice on the chosen issue within the media community.

"Often, a bold idea or an original piece of work will be needed to provide a platform for you to initiate or enter a debate. If you have chosen the right issue it will certainly come round again onto the news agenda and, having staked your claim the first time, you will undoubtedly have another opportunity to make a profile raising contribution. So it is important that you are able to react and retain your position as one of the 'players' in relation to the issue."

Smith cites RICS' Balanced Communities project, running from 2002 until late 2005, as a case in point.

"A report was commissioned and written by an independent, issues-based journalist on increased social division in the urban built environment in US and UK. The issue chosen to focus on was the growth in 'gated communities' in UK.

"A solid media sell-in (briefing journalists, spokespeople, providing fact sheets, figures and case studies) prior to publication and strict embargo resulted in a high level of media coverage, especially in national newspapers. RICS had established itself as first port of call for comment on the issue of gated communities but also on social division in the built environment — new territory for the organisation.

"Coverage across a vast range of media followed from BBC News to Japanese national newspapers. *The Economist* ran the story, giving RICS its fist appearance in that publication. The CEO appeared on Channel 4 prime time documentary devoted to the issue two years later. It's still going on now.

"The breadth of subject gave RICS latitude to present itself as a modern, switched-on organisation, plugged in to the issues of the day. But more importantly, it enabled RICS to fulfil its prime objective, promoting the value and breadth of its members' professional skills — in this case, building functional urban communities in the residential and commercial realm.

"This kind of campaign may appear to be at the luxury end of the market to some. And perhaps risky resource wise. It can be difficult to quantify direct commercial outcomes in relation to the work done. But this is true of much effective PR other than that tightly tied to specific marketing campaigns. The benefits of issue-driven PR have a longer arc but also a more profound payback. It can give an organisation access to a wider sphere of influence.

"Sometimes you can get further when you appear to be selling issues and ideas rather than products and services."

Enter awards

Awards are an extremely important marketing approach for residential developers. Yet there are also numerous awards for the commercial sector that can result in major and valuable press coverage.

However, many awards require a significant amount of time and energy in preparing the submissions; often, media relations agencies are hired solely to ensure entries read well and look professional, and adhere to time scales and entry requirements.

Many residential players regard rewards as important for their brand. "Winning a Bentley award [the most prestigious residential award] in any one year probably doesn't sell any extra properties but it appears on our advertising and contributes to a feeling of achievement in the company and helps build trust in our brand name amongst the public," says one medium-sized residential developer.

But awards are not inexpensive; in addition to the cost of preparing an entry into just one category (some media relations agencies charge up to £2000), the award hosts have perhaps cynically increased the number of categories to attract the maximum number of multiple entries. Then, to add to the cost, the entrants are encouraged to buy tables at central London hotels that are the venues for award ceremonies.

"We spend £20,000 a year on awards. It's probably a luxury we can no longer afford in the current market," admits one large residential developer. He intends to analyse the returns on award entries during 2006 — perhaps a sign that the influence of these events is on the wane in the residential world.

In the commercial sector, there are awards by property magazines, the various professional and regulatory bodies as well as the trade and commercial associations that focus on the sector.

One example that spans both the commercial and residential sectors is what was the PAMADA (Property Advertising Marketing And Design Awards) and will now become the Property Marketing Awards. These awards are managed by The Chartered Surveyors' Company and the main sponsors are *Estates Gazette*.

Graham Chase of Chase & Partners, a retail and leisure property business, is RICS President-elect and is responsible for PAMADA. He says winning awards achieves the same as receiving positive media coverage — it raises standards.

"The PAMADA awards were designed to recognise excellence in property marketing. Achieving coverage for the awards is an integral part of their success. If we raise awareness of high-quality work, then standards improve and, as standards improve, then further awareness is generated and so there is a virtuous circle," he says.

"Whilst *Estates Gazette* supported the awards and provided strong editorial support, when we merge PAMADA with the EG Front Cover Awards to form the new *Estates Gazette* Property Marketing Awards we strengthen that link significantly. This means that there will be even more recognition for property marketing excellence and this should give a real boost to our aim to raise standards in the sector even higher."

There is also a host of category sponsors, including such leading names as MWB group, Rogers Chapman, Rutland Group, Development Securities, British Land, Corporation of London and Allsop & Co. After each award event, *Estates Gazette* features the winners in its news section and includes a special supplement that describes the winning entries.

Depending on how significant the awards are, there are often many opportunities for follow-up interviews and feature articles for the winning entries across a range of media. And while the property sector media will be interested in property awards, a host of regional and local media will be interested in winners within their territory.

Have lunch with journalists

One of the most effective ways of generating quality media coverage is to develop strong relationships with a few key journalists and to make regular contact with them. Whereas in the past, the concept of long, boozy lunches with journalists might have had an element of the truth, in these busy, time-pressured days the media lunch needs a radically different approach.

Lunch provides a social and relatively informal occasion during which you can build and develop a relationship with a journalist. The social element is very important. But the lunch also provides you with an opportunity to learn more about what that particular journalist knows about your organisation and the subjects of interest to him or her and his or her readers. It will also allow them to learn from you; about your role and your organisation, to obtain information about market trends and developments to help them keep up to date on the market and to discuss ideas for future stories or features.

Although the social and relationship benefits — which can and should flow naturally — are important and a valuable outcome of a media lunch, you need to work a little harder. That work starts when you prepare for the lunch appointment.

Preparation — The first thing you must do is familiarise yourself with the magazine (or radio or television programme) that the journalist represents. You may request a media pack (which contains details of the circulation and readership, the forward features list, the advertising rates) or spend some time reviewing the magazine's website. It is also a good idea to read half a dozen or so back copies of the magazine to get a feel for the types of stories covered, the mix of news and features and the general style (eg, informal, gossipy, authoritative, news-oriented, technical, local, global etc).

Once you have a good feel for the nature of the magazine, you should direct your attention to the particular journalist you are meeting. Is he or she a staff or freelance writer? If freelance, who else does he or she write for? If staff, is he or she primarily on the news/reporting side or on the features side? Does he or she have a particular range of topics that he/she writes about? If possible, try to find out in advance what media he or she worked for before and how long he or she has been with them.

Go through the back copies of the magazine you have obtained and review those pieces or articles that the journalist has provided. Note his or her particular style and the range of topics that he or she typically covers.

Journalists will always be pleased to hear you commenting (favourably!) on material they have written and will be impressed you have taken the time and effort to familiarise yourself with them. That's the positive side. The negative side is that if you do not prepare yourself you could find yourself in the embarrassing situation of talking about subjects that have no interest or relevance to the journalist or pitching in ideas that have already been covered. The negative result is that you come across as ill-informed, ill-prepared or just plain arrogant.

Planning — Armed with your understanding of the magazine and the journalist you can start to think about the topics you can explore in more detail and formulating some ideas about what information or ideas you can provide to the journalist at the lunch that they can use immediately.

Your contributed ideas need to take two forms really. Those that might result in some immediate opportunity for your organisation — for example, ideas for articles you might write or news items that you have not provided to any other journalist. And those that are of no direct benefit to you or your organisation but will be extremely valuable to the journalist. In this latter category might be access to information sources or introductions to contacts that you have that might be useful to the journalist.

If it helps, jot down the points you specifically wish to address at the lunch. Remember, once you have settled into the lunch, to raise the different topics. You will be successful if the journalist notes down actions or contacts to pursue after the lunch.

Minding — While it is often easier if your journalist meets alone with your firm's representatives, if your people are inexperienced in working with the media or have not met the particular journalist before, it may be useful to have a PR "minder" present.

Usually, the PR minder will have the advantage of knowing both the firm's people and the journalist and can ease introductions and initial discussions. The minder can ensure the discussion proceeds along the intended lines and prompt the firm's representatives to talk about topics they had prepared. Should a tricky situation arise, the minder can help the firm's representatives manage the situation most effectively or provide them with some thinking time. Having a minder present means that he or she is completely familiar with what was discussed, can identify areas to improve upon for future interviews and make it easier for the minder to do the necessary follow-up work after the lunch or interview.

Follow up — Once the lunch is over, you should write a short e-mail regarding any points that were raised that require further information. For example, send through additional information that you promised, provide a synopsis of an article (see below) that was discussed etc. In order to maintain and build the relationship, you should also set up a system — much as you would with your clients and key referrers — to contact the journalist again at some appropriate point in the future.

Have a journalist to lunch

While the idea of a lunch with a particular journalist is good, most journalists and senior people will be under heavy time pressure, so you need to find other ways than the usual one-to-one lunch or lunch with one journalist and a few members of your organisation.

Some PRs warn against mixing your audiences but there are many successful examples of extremely effective relationship-building and coverage-generating where journalists are just one element of the participants at a lunch or dinner.

For example, increasingly as part of an organisation's overall marketing efforts they might invite a group of clients, potential clients, collaborative organisations, the other guests and the media to join them for a roundtable discussion over a meal or for some form of workshop. While is it important to ensure all the guests are aware of and comfortable with the idea of journalists being present, these events can be extremely valuable to all those involved. There are networking benefits — different people establishing contact with those they have not met before as well as information exchange benefits — discussing issues and comparing views and expertise and hearing about a subject from a number of perspectives can be as useful to your organisation and the other guests and provide the journalists with enough different quotes and ideas for a full and well-balanced article.

The focus of these events may alter significantly. Sometimes, they are positioned as an opportunity mainly for the journalist to meet with and obtain the views from a variety of different people and organisations. Sometimes, they are positioned as a debate or discussion on a particular topic and the media relations element is just one of many. Clear objectives are important and these must be properly communicated to all participants in advance.

While journalists will protect their right to write what they wish without checking material with contributors in advance, for this sort of event — where there are several very senior people present — it may be possible to have some say in the material that results. But the rules of engagement will need to be agreed with the journalist (and the participants) in advance.

Corporate events

The idea of a press launch was mentioned above but there are other types of corporate events where journalists are involved.

An AGM is a time when journalists will be present and will expect detailed financial information and access to senior members of the firm for responses to and comments on a range of specific issues. Financial PR, as we said at the outset, is a specialist media relations function and one that requires skills and knowledge that are beyond the scope of this book — so please seek specialist advice.

Other formal corporate events might be a special periodic briefing where journalists are invited along to a briefing and then one-to-one sessions with senior executives either on their own or as a group.

Another, less formal, type of corporate event is the press party where an organisation invites a large number of the journalists it works with to a purely social occasion where the development of personal relationships is the primary aim.

Successful case studies

Case study: a press party — From the marketing director of a leading surveying practice. "I organised a press party for a major firm of surveyors. They had had rather poor relations with the media in the past and needed to signal a step change in their attitude to dealing with the media. The firm had also been through a number of significant management and strategic changes and wanted to convey these in an informal way. They wanted a foundation on which to start building more productive relationships with the media. Whilst we had a programme of one-to-one sessions with the older, more senior editors it was felt that we needed to get the younger surveyors in the organisation to meet with and start building relationships with the younger reporters in a broad range of property, national, retail, financial and technical media.

We hired the VIP suite of a leading London night club. The first part of the evening was very much like any other reception, there was champagne to drink and canapés to eat. The firm had prepared a press pack containing press releases on five different stories (three different deals, their new website and an expert's opinion on a property fund development) and a host of background material about the firm — as well as photos and biographies of a number of both the senior and junior surveyors present. The press pack was provided as a means of allowing initial conversations to get started — rather than to generate any coverage. By nine o'clock the more senior members of the firm and the more senior editors left, leaving the younger surveyors and the younger reporters to enjoy the rest of the evening drinking, socialising and dancing.

As it happens, the event did generate some positive media coverage immediately after the event but there was a stream of articles and other mentions over the following six months as the various relationships and ideas developed. A less tangible result was a large number of journalists having a very different view of the firm and its people compared to its historic reputation, which made all future communications with journalists a lot easier to manage.

Of course there were one or two instances that would have been best avoided — for example, one rather merry surveyor failing to understand a female journalist's explanation that she worked in news rather than features so she wouldn't be interested in an article (at the post-event briefing we recognised the need to brief all attendees on the different roles of journalists in future) and another young surveyor talking prematurely about one deal (although there was no negative result from this conversation and again, we recognised the need for more in-depth briefing of what could and could not be disclosed at future events). But, overall, the firm was extremely pleased with the event and the journalists who attended felt similarly."

Case study: activity day — A regeneration body organised an activity day for residential property companies and writers (so estate agents, developers and journalists) at a race track, ostensibly as a day of fun and entertainment for those who had already worked or written about the area, but also as an opportunity for networking.

The schedule was packed but each event in it (driving sports cars against the clock, taking a car onto a skid pan, and so on) involved different individuals teaming up; this ensured the ice was broken when two or more people did not know each other.

It provided a rare opportunity for individuals from organisations as diverse as national newspaper contributors, representatives from the Office of the Deputy Prime Minister, local developers and estate agents to talk between sessions in what was a crowded but enjoyable day.

By the end of the morning, conversations were flowing about projects that would not normally be significant enough to be presented to national journalists, and a series of stories flowed from the event.

As with press launches and conferences, there are inherent dangers in having groups of journalists together and many preparations must be made to both maximise the opportunities presented and minimise the risks.

Attend exhibitions

The main purpose of attending an exhibition is to provide an opportunity for members of your organisation to meet with a large number of people from existing and potential client organisations and the various influencers and intermediaries in your market. However, an important element of the overall plan for any exhibition should be media relations and it needs as much planning and preparation as all other aspects of your attendance.

Most major exhibitions will attract a large number and variety of journalists. For this reason, they will have a press office — an area that journalists can use as a base, hold meetings and obtain information about the exhibitors. This means that the very least you should do is to prepare a press pack for your organisation — describing the main products or services you are promoting at the exhibition and providing background information about the people from your organisation that will be present and, most importantly of all, contact details (usually mobile phone numbers) for those people attending.

Some exhibition organisers, and REED is a good example with its annual MIPIM event in Cannes, France will produce a special daily publication for the duration of the exhibition to provide a news channel for everyone attending the conference and exhibition.

For this reason, many organisations choose an exhibition as a good time to release information about new products and services or about major changes in their organisation. Let us consider the merits of this approach.

On the plus side:

- There will be many journalists present who will be available to talk to you and your senior colleagues about the news you are releasing.
- There will be additional channels in which your news can be carried as well as the special exhibition organisers' publications, many publications will devote articles or even special features to the exhibition.
- It is time-effective for you and your colleagues, they are committed to being at the exhibition so they may as well devote some of their time there to meeting with journalists.
- In addition to the time during the opening hours of the exhibition, there will be many opportunities to socialise with journalists out-of-hours.

However, there are also some downsides:

- Your news will be competing with the news from all the other exhibitors, which means it may get less attention than it deserves.
- Very often, the time immediately before and after an exhibition are "slow news" times as people save their news for the exhibition — therefore you are missing a valuable opportunity to get your news to journalists at a time when there is less competition.
- You must ensure that all of the representatives from your organisation are fully briefed and confident on how to work with journalists. It is not unknown for journalists to visit an organisation's stand and start asking whoever is on the stand questions about actively promoted or covertly hidden activities and stories.

It is also important to stress that exhibitions are a time when people relax out of hours and their ability to judge what information may or may not be disclosed to journalists may be affected by tiredness or alcohol. All you can do is warn people to try to behave responsibly and hope that if any journalists do obtain information that they shouldn't that they too will be similarly affected and their ability to use the information impaired. But, then again, we have yet to meet a journalist who can not remember a juicy piece of gossip or an ill-advised comment regardless of their state at the time!

Create "rock stars"

Another powerful strategy for raising the media profile of an organisation is sometimes known as the "rock star" approach. In essence, this involves identifying a few individuals at your organisation who will be at the forefront of the media relations campaign — using their personalities, particular areas of expertise and forthright opinions or expert views to generate coverage.

Obviously, your choice of who your "rock stars" will be will depend on a number of factors — including your overall business and marketing strategy (what elements of your business are you most keen to promote?), the attractiveness of the individuals (both in terms of the weight of their reputations in the property or business marketplace and how photogenic they are) and the personality, skills and confidence of the individuals in dealing with the media.

In effect, you need to build a mini-media relations campaign around each "rock star". You start by building a portfolio — a number of good photos, a compelling CV and a one-page side of bullet points about their chosen subject alongside some pithy quotes. You then need to identify the topics and issues on which they will talk to the media and identify a handful of journalists and magazines where you wish to concentrate your efforts.

You then work hard to get your rock star — armed with the appropriate opinions, comments, views and charisma — in front of the relevant journalists on a regular basis. These people need to be relatively senior and they must be prepared to devote a significant amount of time to working with the media to build relationships, maintain contact and convey interesting opinions and views.

Needless to say, for a rock star to be successful they must be prepared to stick their necks out a bit and say things in a way that others will not dare to emulate. This takes real confidence and nerve and not many are able to do it. But when you look at the few real property sector rock stars, you can immediately see the immense value they bring to their organisations in terms of general profile and reputation terms as well as in specific deals and transactions.

The following example is taken from *Accountancy Age* magazine on 25 August 2005.

Military manoeuvre

On the frontline: Land Securities' new FD will need to draw on all his experience for life in the property fast lane writes Karen Day.

Next month will see Martin Greenslade exchange armoured vehicle manufacturing for retail parks and property as he moves into the FTSE 100 big league to become group finance director of Land Securities.

Greenslade's move from finance director at Alvis to the UK's largest quoted property company is seen as a big leap up the ranks for the 40 year old.

He will swap a traditional military manufacturer, which employed 2,500 staff in Europe, to a company with a complex portfolio of interests including New Scotland Yard and £5.3bn invested in retail property alone in the UK.

Managing the diverse financial interests of Land Securities, which owns or runs 28 shopping centres and 30 retail parks in the UK, will undoubtedly prove a challenge for the Cambridge graduate, especially as he joins just as long standing chairman Peter Birch looks set to exit.

Birch, former chief executive of Abbey National, said Greenslade's "entrepreneurial background" and "successful track record as a finance director" secured the role.

Greenslade's experience in acquisitions and debt restructuring are also likely to have struck a chord with Land Securities' executives.

His entrepreneurial expertise — he founded a corporate finance firm in 1992 — could be stretched over the next 12 months due to the introduction of real estate investment trusts by the government.

These will regulate the tax, investment and finance of the commercial property industry and, because they have the potential to give companies such as Land Securities a greater foothold in residential property, the move could be very advantageous.

Greenslade also joins at a buoyant time for the group, which is a market leader in retail property, the London market and property outsourcing. Despite a £3.2m debt restructuring, which prompted a £155.8m write-off in exceptional costs, its profits for the year ending 2005 was £401m, with a 16.6% increase in the total dividend paid to shareholders.

Greenslade's track record at Alvis, where he oversaw a three-fold increase in its market capitalisation, is impressive, but his five years at the armoured car manufacturer wasn't always an easy ride and will be good preparation for the ups and downs of the property world.

He was forced to oversee a substantial restructuring of Alvis early last year after the government pulled the plug on a potentially lucrative armoured vehicle programme as the export market took a downward turn.

Alvis had acquired military manufacturer Vickers to oversee part of this work, and Greenslade shed significant jobs at its Telford site to keep the company on track. A Norwegian subsidiary was also closed.

His restructuring was evidently successful, with BAE Systems snapping up Alvis by in August 2004.

After managing part of that merger Greenslade moved on, as Alvis was subsumed to create Europe's biggest armoury group.

There are few things in common between military manufacturing and commercial property management, and chartered accountant Greenslade will have to find his feet quickly if he is to successfully steer the group's £10bn combined investment portfolio.

A softer and easier way to build an individual's media profile is by them adopting a spokesperson role for a particular professional, trade, business or community association. This way, they develop their personal profile alongside that of the group they speak for, and your organisation receives a subtle "halo" effect from supporting the individual in this outside-of-their-main-job role.

A good example here is the increased profile achieved by Ian Coull, who left Sainsbury's in 2002 to become chief executive of Slough Estates and then became vice-president of the British Property Federation in 2004 and president in 2005.

Another approach is to encourage an individual to develop relationships and provide articles regularly to a magazine and, in time, to negotiate a regular slot or column where they become a regular and fixed feature of the magazine. Some magazines run panels of experts who are available to answer questions of a specific nature on behalf of their readers, other magazines have regular slots that they rotate among a small number of regular "guest" writers.

The only danger with a regular slot is either that after some initial enthusiasm, the contributor might run out of energy and/or new topics for the column and this results in poor relations with the magazine that now has readers expecting to see quality material on a regular basis. So choose carefully who you select to suggest for a regular slot.

One final word of caution is required on rock stars. Nick Mattison, director of Mattison Public Relations, says the property industry has so many strong, colourful characters that many prefer to be seen as celebrities rather than professionals, with perhaps too little substance and too little by way of genuine things to say.

"I remember having to spend hours responding to an excited property marketing person who thought one of her directors should get national coverage because he was in a charity cycle ride and was the spitting image of a notorious radio and television celebrity. It was difficult to generate similar excitement with heavyweight stories or comments," he says.

"On another occasion, and with a different client, we thought we had got the message through after a two-hour workshop. At the end, one of the participants thought he had a great idea for us. What's that? "We've just been voted the second best employer in Norwich — surely we could do something with that.'"

Hold the front page!

Write feature articles

As described earlier in the book, most media will have a balance between news and feature materials. If your organisation has little news that it can use to generate coverage, then feature writing provides a valuable alternative.

However, as writing a press release is different from other types of marketing or technical writing, so is writing an article. And, to be truthful, there are relatively few people in the property sector who have the relevant skills and experience in writing really interesting articles with a broad appeal.

Therefore, the people who will byline the articles will probably need some help from either their in-house or agency media relations consultants or from a specially commissioned ghost writer. It is worth noting, at this point, that some staff writers and many freelance writers will be happy to assist your firm with writing projects for a fee.

A word of warning on articles. Often, an enthusiastic property professional might take the time and trouble to write an article on a topical or technical subject and then expect a PR professional to "place" it with the chosen medium. This rarely works out happily.

A much better approach is to identify the topic for the article and prepare a synopsis so that it can be pitched in to a magazine or paper and that a firm commitment (or commission) is obtained before the author commits the time and trouble to writing the piece.

If you speak to the features editor before writing in earnest begins, you will get a clear description of what will be contained in the article, exactly how long it should be (there is a world of difference between a 500 word article and a 3000-word article) and when it must be submitted (the deadline — so-called because if you miss it your media career will be dead or will be shortly after letting down the editor/journalist involved).

Pitching an idea

The first step is to identify the suitable author, prepare a synopsis of the article you wish to write (see below) and pitch it to a newspaper or magazine. As we have seen from earlier in this book, this process can be long and sometimes unrewarding. It is perhaps better handled by someone other than the proposed author — for example, the in-house PRO or PR agency account executive. They can then act as the intermediary between the magazine and the author and ensure that the media's needs are understood and fulfilled in the article.

You might pitch the article in as part of a regular discussion with a known journalist. You might use one or two different article ideas to initiate a conversation with a journalist that you have not dealt with before. You might simply e-mail the synopsis to a journalist with whom you communicate regularly.

You can also combine your rock star strategy with an articles programme — with regular releases summarising the expert's view or opinion on a particular subject and offering these as the basis for an article.

Preparing a synopsis

Your synopsis should be no more than a single side of A4 paper and contain roughly the following:

- Name/organisation/date/contact details.
- Potential title of article.
- Broad description of what the article is about: Short paragraph describing the purpose and content of the article.
- Context: Why the article is relevant/topical now — a legal development, an imminent market change, a recent transaction etc.
- Relevance to the readership: Why the readership of this particular magazine should be interested.
- Main content: No more than six bullet points on the main contents/messages.
- Likely length: (Indicate whether there is a minimum amount of words and also the ideal amount).
- Background of the author: Reasons why the author is best positioned to write the piece, key credibility statements (eg, qualifications, experience, positions of responsibility, acknowledged expertise etc).

Using forward feature lists

Another approach is to review the forward features list of the relevant media and pitch the synopsis in as part of that feature. Another word of warning here — features are designed to generate advertising revenue (see advertorial below) and so sometimes they will only wish for articles from those who back up their article with advertising.

However, special features and supplements are usually prepared well in advance of their publication date. Sometimes, this means getting your synopses to the magazine up to six months in advance of the publication date. For popular features, there will be a great deal of competition from other people who are keen to have material included. The other point to consider is that often the special features and supplements are outsourced by the magazine to be written by freelance journalists or contractors. This means that they may not know your organisation as well as the regular reporters at the magazine and that they will be less interested in a long-term relationship and more focused on what they specifically need for that particular feature or supplement.

Learn from pitch discussions

Whether or not that particular article is accepted, you should pay careful attention to the conversation with the journalist about it as you will learn a lot about that medium's attitude to contributed articles. Some media have set policies about who they will accept material from. Some will be keen to accept material but reserve the right to alter it significantly. Most will want to have the copyright of the article to use in online versions. Some might indicate that they are happy to have a byline crediting the individual — but not the organisation. They will all stress that the article must be just that — an article — and not a marketing/promotional piece about either the writer or his or her organisation.

But in addition to the general policies and approaches towards features, the discussion will reveal what types of topics *are* of interest to them and their general requirements for contributed articles. A well-organised media relations professional will keep detailed records of the needs and preferences of both the key publications and the individual editors and reporters so that they build a good knowledge base of who is likely to be interested and in what material, which makes future article pitches more accurate and likely to be successful.

Writing and submitting the article

Once you have a firm commitment from a magazine that they will publish your article, you must write it. The deadline looms and you put off the dreaded task of starting to write.

How people approach the writing task varies hugely. Some people simply start typing — a sort of stream of consciousness approach — and, after a short break, go back and structure, edit and refine the material. Others will spend time identifying the overall structure of the article, listing out the main points to be conveyed and approach the writing in a systematic and structured way.

Those who have difficulty articulating their ideas on paper may prefer to tackle the task by speaking the words out loud and hoping that someone or something (dictating machines are useful here) will capture their words, which they can then edit into an article at a later date. Another approach — again, this is helpful for people who are brilliant at their subjects but short of time and interest in writing — is to have a media relations professional of some kind "interview" them — asking questions, probing responses, asking for examples, challenging etc — and write down what is discussed. The media relations professional (or ghost writer) can prepare a first draft of an article for the author to edit and refine as required.

It is always advisable to wait a day or two (if deadlines permit) after writing an article and then come back with fresh eyes to check it for a) flow and style b) fit with what was agreed in advance and c) the word count. While no one will get too upset if you submit 1500 words for a 1400-word article, there will be major tantrums if you submit 3000. It also helps to ask someone unrelated to the article to read through a draft and make comments and observations that will enable it to be further refined.

You must take care with any material you incorporate from other sources — so acknowledge research and other sources. If you have mentioned clients or other organisations in the article, make sure that you have their permission to do so if this is required. Sometimes obtaining permission is not required but it is a polite and responsible thing to do. If you can provide photos and illustrations this will be of considerable assistance but check the position on copyright.

Once the article is submitted the editors may ask a few questions — to confirm spellings or facts or to check that contributed material has the necessary permissions. It is also common for the publication to send a proof of the article for the author to read and approve. Please bear in mind that such proofs must be turned around very quickly (usually within 24 hours) and that they do not expect you to be making changes to the content — only corrections.

A final word of warning — although it may be tempting to try and place a particularly good article with more than one newspaper or magazine, this is to be avoided at all costs. You may destroy a carefully nurtured relationship with a magazine if they find that the article they have just published by you has already appeared in another magazine.

Case study — SJBerwin

Jon Vivian, a real estate partner at property law firm SJBerwin and his occasional column in *Property Week*:

"The property press is a very different animal to the legal press. It is difficult for us as lawyers to convey information about deals to the property press — clients seem to accept that the agents will talk to the media about deals and do not mind — but they do not expect their lawyers to be discussing their business. As a result, we often can only announce deals when they have already been reported in the property press. We certainly could never leak them! The legal press are more intrusive to law firms. They are more interested in what the firm is doing than the deal or property news *per se*. They are more likely to pick up a bit of information about the firm and blow it up into a major story.

"Property law can be very complicated. Very often the property media is not interested in the intricacies — just the overall impact of a recent change and the implications for investors. Not everyone finds it easy to write from that perspective."

Vivian describes how he approaches writing his column in a bid to attract the maximum audience from the *Property Week* readership:

"I create a picture in my mind of a surveyor — assuming a reasonable level of intelligence and awareness of legal issues and I try to imagine what he or she is interested in. Then I write from his or her perspective rather than from my technical, lawyer perspective. You have to keep your target audience in mind when you write. It is a real challenge to identify topical things to write about on a regular basis. And it is hard to convey sometimes very complex issues in just 700 or 800 words. However, I quite enjoy the challenge of taking a couple of hours out of a busy work schedule to identify an issue, do any relevant research and write in a way that is compelling."

Case study — haysmacintyre

Graham Elliott, a VAT partner at haysmacintyre, uses the media extensively to inform real estate industry players of legal decisions and changes in official policy.

"Nothing replaces phoning or e-mailing existing clients and contacts but this does not address the real estate companies to which I am not connected. Almost all communication with them is via the media, so the sooner I can inform the media in a way that is clear as to the importance and consequences of a change, the quicker every interested party will find out, from me. This means that it is always better to have established media contacts, where trust has built up, so that I can place a piece, or contribute some comments, as the issue unfolds.

"I convey a great deal of information to new potential clients and contacts through a range of important media outlets. Even my existing clients pick up on these from time to time, and it prompts them to get in touch to explore how the reported issue impacts on their latest plans.

"One time, a client sent me a copy of my latest column in *Accountancy* magazine, without noticing that I had written it, and asking for my opinion! Other clients sometimes comment that they like the fact that 'their VAT advisor' is regularly featured in respected publications. Once a practitioner from another accountancy practice phoned me for advice after reading my piece on property conversions in *Taxation* magazine (it only took five minutes, so I didn't charge him)! Of course, the publication you choose to be featured in says a great deal about the kind of practitioner you are, so it is wise to save yourself for the publications that have maximum marketplace respect, and which fit with your style or approach.

"I have written for *Accountancy* magazine on a monthly basis for 10 years. This publication fits the image I would like to portray of myself. I write frequently for *Charity Finance* magazine, the leading technical magazine in its sector, and latterly for *Estates Gazette*, again, a market leader in its sector, and, just as important, a publication that people trust.

"Having written for many years for the financial media, and having contributed videos for training channels, I have a reasonable feeling for the traits that editors require in their contributors.

"There are the obvious points about reliability. The contributor must contribute on time as agreed. He must be technically reliable. These are the starting points. But an editor wants more than that. He or she wants copy that flows and will not deter less knowledgeable and more generalist readers. If it is written for a particular market segment, such as real estate, the contributor must talk the language of the segment, not the in-talk of the technical subject. Wherever possible the subject must be brought alive by the judicious addition of some opinion and observation, as long as these are sensible and not sensationalist. If a development, such as a case decision, seems questionable, then it is OK to say so. Allowing some part of your own philosophy and even personality to come through in the piece helps to humanise the process for the reader, and this makes reading and remembering that much more easy."

The dangerous world of advertorial

At the beginning of the book, we described the difference between media relations generated editorial and paid-for advertising.

While most magazines will have separate teams dealing with editorial and advertising, there is inevitably a link between them. It is important that you understand the relationship between them in any particular medium.

We have mentioned the advertising objectives of features. The advertising teams will be keen to ensure that major advertisers with their features are given a fair crack at the editorial content. While the editorial teams might resist this in principle, commercial success means that it is often a reality.

However, for some magazines the main way they generate advertising income is through advertorial. This is where the author/advertiser writes an article but has to pay for the privilege of having it published. So it is really an "advertising feature" rather than an article. The good news in these situations is that, as author, you really do have editorial control — what you write in an article is what will appear. The downside is that most of the readers will know that the articles have been paid for and are closer to advertising than editorial and give it the same attention and credibility.

This advertorial approach is prevalent with the regional business magazines and if your target audience is business tenants in a particular region then you will have little choice but to accept the price. This approach is also common in many mainstream consumer newspapers and magazines, so the residential property market is likely to be more familiar with the costs and implications of advertorial approaches.

Other approaches

In this chapter, we have reviewed some of the most common approaches to proactive PR but there are hundreds of others. In some respects, the creative thinking and ideas behind some of the best approaches are what you pay good in-house and external media relations consultants for. It is more than our lives are worth to give away all of their secrets!

However, in this section we review some of the other proactive media relations approaches that have been used successfully in both the commercial and residential property sectors. As we would like to include more examples in future editions of the book, please submit your ideas to the editors.

PR stunts

A PR stunt can take many forms. Typically, they centre on creating a really good and unusual photo opportunity for the media. There are numerous examples within the property sector — for example, senior people abseiling down the outside of a new building, staff dressing up in unusual costumes and getting up to surprising antics among members of the public in a shopping mall or putting high-profile people in unusual locations.

While care must be taken to ensure that the stunt doesn't backfire (eg, bad weather preventing an out-of-doors activity taking place, the senior people making complete fools of themselves doing something completely inappropriate — think of Tony Blair as a rock guitarist), these stunts often generate a lot of good (photographic) coverage for your firm, development or cause.

Human interest — A day in the life of ...

Many magazines have a regular slot that focuses on the human side or people's lives beyond work. Sometimes these are full-blown profile articles, sometime they are "A day in the life of ..." pieces and other times they are in the form of "30-second responses to 10 standard questions".

It is worth finding out if any of your directors or staff have hobbies or out-of-work activities that are unusual or interesting, as this can be an added bonus to those who produce the profile sections.

Celebrities

Particularly in the residential market, celebrities can be used to attract media interest, encourage the media to attend launches and to form the centrepiece of photographs that also convey some element of your organisation or developments.

The cost can be prohibitive though — the best celebrities will have hefty appearance fees. You must take care to ensure the chosen celebrity is fully briefed about the nature of your organisation and the messages you are promoting so that they act as credible ambassadors for your project. There's nothing worse than your celebrity mispronouncing the name of your organisation, one of its directors or the development being promoted to turn a media opportunity into a media disaster.

Celebrities are also difficult to control — they may arrive late or in inappropriate attire or say the wrong thing. There is also the risk that some news element of the celebrity will eclipse your organisation's story.

Another area to take care of is how your representatives will look in photos with a celebrity. Think glamorous young pop star standing next to old, grey-suited executive ...

Fundraising and charity events

If people within your organisation undertake some sporting or other challenge in order to raise money for charity, then there may be an opportunity to generate some positive media coverage. The upside is that your organisation and the charity will benefit. There will be particular interest within the local media if the charity or cause has some local angle.

However, care should be taken. Another photo of a group of people running, swimming or cycling or of a grey-suited executive handing over a large cheque is unlikely to generate much interest. So you will need to find a more exciting angle — perhaps one member of the team has some interesting personal story relating to the challenge or the charity, perhaps your organisation has some special link with the charity?

You may need to think laterally to identify a suitable story or unusual photo to get published. And you should take care not to be seen to be only undertaking the fundraising or charitable event for publicity purposes.

Some of the best fundraising exercises are the result of many, many years of effort and commitment — and only generate a very occasional piece of publicity. It is important to remember in these situations that the fundraising or the cause is the central objective and any coverage is a welcome by-product.

Internet and weblogs

The whole area of electronic communication is developing very fast and there are increasing and new opportunities to generate media coverage using this medium. One of interesting recent developments has been that of web seminars and weblogs. The web seminar provides a forum where a senior executive — anywhere in the world — can provide a presentation and then field questions to a group of seminar attendees across a wide geographical area. From a control point of view, this method of communicating with a group of journalists offers attractive benefits — the questions are posed by the participants and hidden from the other delegates. The moderator can decide which questions to field and can control the onscreen and audio questions and comments from the participants.

Weblogs, a kind of open online diary, provide another interesting source of potential media coverage if the individual or situation is sufficiently interesting. However, you should bear in mind that it is public (unless in a password-controlled area) and most media will require some form of exclusivity before they use any material.

Competitions

While the consumer and more sophisticated media may see competitions as a major revenue stream and therefore treat competitions as another form of advertising, other media may respond well to the offer of a carefully thought through competition. There needs to be some attraction and the obvious one is a really valuable prize — which will cost you money. But competitions without a significant prize — but with plenty of community liaison and direct community interest — may be of interest to local media. A good example is the type of competitions that require participation by schools or other community groups that focus on some local site, development or facility.

Competitions that run over a period of time offer several opportunities for coverage — as the competition is launched, to review entries and progress during the competition, to announce the winners and then to write in-depth features on the winning entries and/or how those winning entries are used.

Radio and television interviews

Much of the information contained in the book generally relates to the broadcast media as much as the print media. However, there are some significant differences in which stories are suitable for and how they are pitched into and produced for the broadcast media. This is really an area where you should obtain some expert help.

However, the key things to remember are:

- You can convey only a very small amount of information on radio or television. Much less than in print. Therefore, you must focus on the key messages.
- For television, great care must be taken of the visual impression created. This extends beyond the physical appearance and attire of the person to their non-verbal communication (body language) and gestures and facial expressions.
- To be interesting — and to allow for editing — each statement or sentence must be short and self-contained.
- If you are part of a panel or discussion, you must be prepared to be eclipsed, interrupted or challenged by the other panel members, who may have very different agendas, perspectives or views of professional/appropriate behaviour.

- A live broadcast is extremely nerve-racking for those who are speaking or being interviewed and there are many variables that can affect the quality.

A good tip is to summarise a) the key elements of the story b) the relevance to the target audience c) the name, position and organisation of the individual and d) provide a short biography of the individual in large type on a single sheet of paper. This is of considerable use to the interviewer and is likely to result in him or her getting the names right and using them during the broadcast.

More so than in non-recorded media interviews, the key will be to prepare and rehearse as much as possible so that when on air/camera the individual comes across as friendly and relaxed. The need for specialist "front of camera" training is mentioned on p130 above.

Syndicated radio tapes

As an alternative to live interviews, there are agencies that will help you script, prepare and produce a series of short (eg, 30, 60, 180 and 300-second versions) radio pieces, which are then sent out to a number of radio stations on a syndicated basis. This type of approach can be targeted towards specific regions.

Building relationships with journalists

While this chapter has considered a variety of methods to generating proactive media relations coverage, underlying all approaches and methods is the need to develop a good understanding of and solid working relationships with the most important journalists for your organisation.

Building relationships cannot be done overnight. It takes a serious commitment to ongoing communications over a substantial period of time. The journalist needs to learn that your organisation will provide good quality information, identify stories and material that will help him or her provide strong (and exclusive) pieces and that you will provide fast access to senior people within your organisation who will say things that they will want to print.

However, the relationships can be a little one-way. You may have to invest a lot of time and energy in a journalist relationship before any coverage results. And, at the end of the day, regardless of how well you get on with the journalist, they can only publish stories about your organisation if you provide good stories and relevant information.

Similarly, while having a long-standing relationship might help you gain some sympathy for conveying your perspective on a difficult or negative story, it is unlikely that it will insulate you entirely from the risk of negative coverage. And while it can take years to establish a good relationship with a journalist, it can take just one unfortunate incident to destroy it overnight.

You must also remember that journalists tend to move up and away from your target media. Sometimes they move to other publications that are relevant for your organisation, other times they move to media that are not. Then you have to start the process all over again with their replacements. The time required to maintain strong relationships with a number of journalists across the national, trade, consumer, regional and local media is one of the main reasons why organisations commission PR agencies to manage those relationships for them on their behalf.

Case study — Developing long-term relationships with journalists

Director of marketing at a leading property law firm — "Journalists move around a lot — and sometimes they may have links with other media that might be valuable to you. Other times, they will move to media that are outside the scope of interest to you — but it is helpful to stay in touch because they may return, one day, to a medium that is relevant to you. So consider relationships with journalists on a long-term basis — not purely for the short-term potential gain.

It is also important to recognise that new/junior reporters progress through the ranks very quickly. There is one very good example that springs to mind. A young rooky reporter — knowing absolutely nothing about the property sector — joined a leading legal magazine. While being bright and personable, it was clear that his property knowledge was very limited. I persuaded some of the most senior property partners at the firm to spend a lot of time with this young reporter to help him build and grow his knowledge and contacts within the property sector. I explained that there would be no short-term return for this significant investment of their time. Reluctantly they agreed. In addition to almost a full day's initial briefing involving around six partners, those same partners provided many additional hours of time responding to telephone calls to help the young journalist understand the intricacies and implications of the stories he was writing about.

Over the next 12 months, the partners and the firm did receive some positive coverage — their comments and opinions on developments that the journalist wrote about in the course of his work. Then the young journalist moved to another magazine in the sector — a much more upmarket and highly respected journal with a much greater cachet. Having established such a strong relationship with the partners at the firm, the reporter — who is now among the ranks of the senior editorial team — continued to return to those same partners for input, assistance and comments. The partners were delighted as this was a magazine that had formerly been beyond their grasp. The partners continued to invest time and energy in maintaining regular contact with the journalist — providing ideas and information that had no direct benefit to them or the firm in terms of potential coverage.

After about three years though, things took a really interesting development, the young journalist became the editor for the most prestigious and important (to the market) publication in the sector. The new editor was courted hard by a multitude of other firms trying to win his time and attention. Yet he relied on those people who had helped him early in his career — he trusted and liked them, had very well-established relationships with them — both for business purposes and on the informal, social side. So the partners who had invested all the time over the past five years or so were those he turned to in his new role as editor. It took much time and energy over those years, but the partners started to really reap the benefits some five or six years after their investment."

Measuring Effectiveness and Results

Why measure media relations effectiveness?

The old adage "What gets measured, gets managed" is true. Without measurement, there can be lots of media relations activity but without the necessary management to ensure desired results are achieved.

It is important to measure the inputs, outputs and impacts of all the activities in both profit and non-profit organisations. So measuring the effectiveness of your media relations is no exception.

In addition to assessing performance and identifying ways in which to improve that performance, businesses must also measure the return on investment. Media relations can be expensive with in-house staff salaries, external PR consultancy costs, internal information systems and processes, media events and publications and most of all senior executive time all taking considerable and often long-term investment. So what's the return?

There are many cynics around. They expect to see hard results and payback for any investment. Media relations, being a straight cost to the business, is no exception. Without compelling results, usually shown as hard figures, they can effectively persuade the "undecideds", overpower supporters and pull the plug on media relations activities.

But it is the experience of the authors of this book that within the property industry there are few examples of organisations with sophisticated media relations measurement systems. We hope that if media relations can be properly managed and measured, it will encourage more investment and better coverage — which will benefit everyone in the property industry in the long run.

Difficulties in measuring media relations effectiveness

While it is important to have measures for all marketing and business development activity, it is not as straightforward as one would expect.

The 2003 Department of Trade and Industry (DTI) study of the PR industry (*Unlocking the Potential of Public Relations*) identified the primary purposes of PR as being reputation management, awareness raising, stakeholder relationship building and crisis and issues management. These are complex, long-term benefits that cannot be reduced to a financial equation.

Media relations is just one aspect of reputation management

In the area of reputation management — one element of which is media relations — it is difficult to develop measures of a baseline or of progress against that baseline as something as nebulous as "our good reputation" or in marketing speak "our brand". Furthermore, the reputation is created and perpetuated by many factors: how staff answer calls or behave at meetings and at networking events, the nature of the material you mail out, your advertising messages, your website, your show suites and even how your accounts team handles enquiries. So how do you isolate and measure the contribution of media relations?

Measuring your reputation is different to measuring your media relations process and results. Your reputation is the combination of the views of your organisation from a number of audiences whereas your media relations reputation is a measure of how the press and media see you, and how they portray you and your organisation in its pages.

Long time frames

Very often, it takes many months and sometimes years of effort to create a measurable change in your media profile. And for those with an established profile there may be a huge amount of effort invested in maintaining the profile that is established with no discernible change. This makes a reliance of monthly or quarterly results intrinsically suspect. How can you monitor a change that takes place gradually over a number of years? How can you measure maintenance of an existing profile?

Subjective

Often, the impact of media relations is subjective. Directors of the business may receive informal feedback from a number of channels about their reaction to a firm's media relations efforts. How do you compare and measure the impact of, say, a senior director of a key client commenting favourably about your firm's recent article in a national newspaper and the positive impression created among a large number of students who are considering applying to your organisation? What about the effect of a key piece of coverage on your share price? And think about the damage done by the careless speak of "crap" by the chief executive of Ratners all those years ago — no media relations programme will ever be able to undo or repair such damage.

Too little exposure and profile

With media relations in smaller organisations or those who have only recently started to actively pursue media relations activities, there is often so little efforts that measurement is not necessary as you can clearly remember all successful (or unsuccessful) results from your media coverage. In some respects, it is easier to measure from a cold start than it is to measure an improvement on an existing media relations programme.

Measuring media relations effectiveness

So how do we try to establish appropriate systems? The experts recognise that there is no one simple answer.

Dr Walter K Lindenmann, senior vice-president and director of research at Ketchum Public Relations at the national conference of the IPR said "Keep in mind there is no one, simple all-encompassing tool or technique that can be relied upon to evaluate PR effectiveness. Usually, a combination of different measurement techniques is needed".

- Set measurable goals.
- Measure outputs.
- Measure outcomes.
- Link all outcomes to business goals, objectives and strategies.

Set clear goals

As with all marketing, the key to measuring media relations success is to set SMART (Specific Measurable Achievable Realistic and Time-specific) objectives. This way, you can see how well and how quickly you are moving towards your goal.

Setting clear objectives will also make it easier for your in-house and agency professionals to focus their efforts and meet your expectations. Agreeing at the outset of a campaign what you wish to achieve and in what time frame will mean that the right resources are directed at the right activities to achieve the right results.

But, as mentioned above, if media relations takes time to have an impact, what measures can you set to help you measure progress at the start of the programme?

Figure 6.1

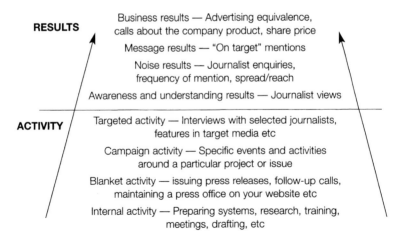

Media relations pipeline

RESULTS
Business results — Advertising equivalence, calls about the company product, share price
Message results — "On target" mentions
Noise results — Journalist enquiries, frequency of mention, spread/reach
Awareness and understanding results — Journalist views

ACTIVITY
Targeted activity — Interviews with selected journalists, features in target media etc
Campaign activity — Specific events and activities around a particular project or issue
Blanket activity — issuing press releases, follow-up calls, maintaining a press office on your website etc
Internal activity — Preparing systems, research, training, meetings, drafting, etc

Source: Kim Tasso

One response is to measure both the inputs and process as well as the outputs and results. You can certainly measure the levels of activities even if the results are some way down the line. Measuring media relations activity and results can be see much like a sales pipeline — you need a great deal of downstream effort in order to achieve a small (but significant) upstream result. Measuring the different stages of the "pipeline" can help you identify problem areas.

Measuring the process

Time input — If you are managing an in-house team, you might consider measuring how much time they spend with each department, team, development or project. PR consultancies typically charge on a day rate, so will need to account for the two or five days a month that they spend working on your behalf. If you wish to monitor your PR agency in this way, you will need to agree what information you require from them and in what format in your brief at the outset.

Internal interviews — At the outset of a media relations programme, you might wish to measure how many people within the organisation have met with and talked to the PR professionals about how the media relations programme will be set up and operated. As mentioned in the section above, good internal communication is vital to generate a flow of information that can be used in media relations, ensuring that your media relations people are spending enough time with the right people internally can be a useful measure.

Training — Earlier, we mentioned the importance of ensuring that various internal communications and training programmes were implemented, to help people understand what is involved in media relations and their contribution. So measuring the number of hours of training or the number of people attending workshops and training sessions can be useful in large-scale programmes.

Press releases — In a news-oriented environment, for example, where you have many deals or transactions to report you might wish to measure how many press releases are written and distributed each month. This measure will not work in an environment where you are reliant on feature articles and expert opinion though.

Media interviews — Where the emphasis is on establishing mutual understanding and relationships with journalists, you might wish to measure how many interactions — whether these are telephone conversations or face-to-face meetings with targeted reporters — take place. Ensure that you have indicated which journalists or media are important, otherwise you may find that your PR professionals are setting up meetings with people whom they already know rather than those who you wish to get to know.

Measuring the results

Over time, you should start to see your media relations generate results and there are a number of hard measures that can be used to assess progress.

Hard measures

Journalist enquiries — Measures of calls into a central point, such as an internal press officer or to a nominated account executive at a PR agency, will provide an indication of the extent to which your activity is generating interest from the media. Some organisations measure the number of calls from those journalists to which press releases are sent or to those journalists on a target list.

Web visits — If you have a dedicated press area of your website, containing background information about the organisation and key products or services, up-to-date statistical information, biographies of key executives, archives of press releases etc, then you can measure the number of visits to this area. However, it is almost impossible to determine whether these visits are from genuine journalists or simply from people — perhaps potential employees or clients — seeking information about your organisation. However, they indicate a flow of enquiries that can be measured over time.

Mentions/cuttings — A common but crude measure is how often your organisation or project is mentioned in the press. If it is volume you seek, then this may suffice. However, you will need to check that you have the necessary systems in place to be alerted to mentions as it is rare for organisations to have subscriptions to all the media where their organisation may be mentioned. These days there are a variety of agencies (see the information section in chapter 7) that can be paid to "read" your target media (either advise them of the names of the relevant local, regional, trade/technical, consumer and national media you are targeting or send them a list of the media to whom you distribute news). These agencies can also extend their monitoring to the broadcast media and provide you with tapes or CDs containing the relevant radio or television broadcasts for your organisation. More recently, we have seen the advent of electronic monitoring services that monitor the internet for mentions of your organisation. Magenta is one such example.

Column inches — This established method of measurement considers the overall size of the mentions about your organisation by calculating the length of the piece multiplied by the size of the column. This gives a hard measure of the amount of coverage — but not its quality.

Advertising value equivalence (AVE) — This is one for the accountants among those reviewing media relations progress. In essence, it takes the column inches calculation mentioned above and multiples this against the rate card cost of what equivalent advertising would cost. This method allows senior decision makers to have a financial basis on which to evaluate press coverage that they can compare against other marketing investments (notably advertising expenditure). It also moves media relations closer to allowing some form of Return on Investment (ROI) analysis.

Size or importance of mentions — More sophisticated analysis will consider the size of the mention received — for example, a short quote from an executive or an in-depth profile and whether there is a photograph or table of information included as well as the position of the mention. So, for example, a higher value is placed on a piece on the front page of a local newspaper compared to the appointments section of a trade/technical magazine. Some organisations apply a simple weighting scheme to such mentions so that an approximate score is achieved for the type of coverage achieved.

Share of voice — The above measures fail to take into account the levels of coverage achieved by your competitors, so some measurement systems will attempt to track your organisation's coverage against a group of nominated competitors so you can track your organisation's "share of voice". This type of analysis is complicated and expensive, as it requires the same analysis across a number of organisations rather than just your own.

Enquiries — Some campaigns are designed to generate enquiries — generally from the public (such as for information packs or viewing details) but sometimes from businesses (requests for copies of research). A specific telephone number or micro website is assigned to the campaign and promoted through the media relations programme so that the number of enquiries generated from media relations can be measured separately from those resulting from direct mail or other marketing methods. There have been examples of people achieving enough enquiries to achieve 100% sales or lettings for developments through targeted and creative media relations campaigns.

Share price — For publicly listed companies, media coverage can have a direct impact on the views of investment analysts and subsequently the share price. However, it is hard to separate out the effect of the results themselves from the nature of the press coverage that it generates. More often, the share price is seen to be significantly influenced by the media relations effort and is one of the more subjective methods of media relations effectiveness.

Other systems — Many PR agencies have invested time and money in developing elaborate systems to measure media relations effectiveness in other ways — drawing on a combination of the above approaches. However, it has to be said that these people have a vested interest in showing positive results. Furthermore, the cost of operating some of these systems is almost as much as operating the media relations programme in the first place. So, unless there is a really significant spend and reliance on media relations in your organisation, you are advised to keep things simple (and cost effective).

Soft measures

Internal perception — Very often (and despite what elaborate hard measurement systems might say to the contrary), one of the most important measures is how people within the organisation feel about the effectiveness. Often, senior decision makers will not take a huge amount of interest in media relations measures unless they feel that there is a problem — low profile, incorrect image and so on. This internal perception measure is also important among the other staff too, particularly if your organisation comprises a large number of staff who are in daily contact with clients or customers. It can be quite quick and inexpensive to set up a regular electronic poll, through e-mails or the organisation's intranet, to see how people throughout the organisation feel about the firm's media image. The main concern with this is that the perception can be based on other aspects of the marketing mix — for example, advertising or sponsorship or direct mail — as well as media relations.

Press perceptions audit — A good way to start a media relations programme is with an audit of the views of the target journalists to assess their levels of awareness and understanding of your organisation and to see how your organisation is viewed. PR professionals advise organisations to do this at the outset of a media relations campaign so that activities and messages can be focused correctly. However, this sort of audit can become part of a regular assessment, with journalists' graded ratings across a number of areas and issues being used as a helpful measurement tool.

Tracking studies — Another important audience to monitor is your target markets of customers or clients. Often linked in with brand awareness and perception studies, you can use research in this area to measure the impact of media relations upon a target audience. These exercises can be costly but usually involve some form of questionnaire or interview before a campaign is undertaken and a further study (ideally among the same or a comparable group of the target audience) at regular intervals through the campaign and at the end. As with many measurement methods, it is difficult to tease out the impact of media relations from other marketing activities if it is part of an integrated campaign.

Tone of message coverage — This builds on the number of mentions or column inches measures but takes into account whether the tone of the coverage was positive/favourable to the organisation, neutral or negative. This approach deals effectively with the problem of generating lots of coverage but of the wrong type (ie, negative). Needless to say, these measures will include a subjective judgment and what one person sees as neutral another may see as negative. This can be overcome by agreeing in advance what will be seen as positive and negative and having a small panel of people assessing the borderline cases. Undertaken well, this type of measure can become as "hard" as those above.

"On message" coverage — Other analyses might consider the extent to which the mentions include key phrases or statements from the organisation — its strategic or positioning messages. Similarly, mentions might be measured in terms of the extent to which they are positive, negative or neutral about the organisation. However, the subjective element on the extent to which a piece of coverage is "on message" can be difficult and journalists have a knack of reporting both favourably and negatively on different aspects within the same mention.

Awards — Although not a direct measure, winning an industry award — such as the Chartered Institute of Public Relations coveted Sword of Excellence — can be an indication that you have achieved outstanding results with your media coverage. However, it takes time and energy to prepare the entry submissions and your organisation if small or at the start of its media relations development might not stand a fair chance against larger organisations with more experience and resources to invest.

For those who wish to see more cutting-edge developments in the measurement of media relations effectiveness should refer to *Best Practice in the measurement and reporting of PR and ROI — A research study conducted for the Institute of Public Relations and The Communications Directors' Forum*, which was undertaken by Metrica Research in February 2004.

Case study — Supporting planning applications

"The council bought a WW2 airfield and had plans to build a business park on it based on a very successful one in Philadelphia. It was environmentally friendly and would also support domestic housing. The council said there could be no expansion so we had two weeks to overturn the planning vote — we took the local press at two days' notice to get them together with the leader of the council. We did solid PR editorial when we got home and we won the vote by three, which led to a saving of £28m".

Reported in "Best Practice in the measurement and reporting of PR and ROI — A research study conducted for the Institute of Public Relations and The Communications Directors' Forum" by Metrica Research February 2004.

Case study — Measuring media relations effectiveness at the RICS

Sean Tompkins, executive director of brand and membership of the RICS (supported by a team of 10 marketers, four press officers, four web managers and 23 customer service staff) says: "Two years ago, RICS received 40 mentions in the national press each month — mainly to do with residential housing — that figure today is now over 100 — and with only one additional media person. There are also 200–300 regional press mentions of RICS and over 20 in non-UK media every month. In the third quarter 2003 RICS was the highest quoted professional body. This means that around 80 million people will be exposed to RICS each month. And two years ago, we had only half of the million website hits we receive each year now. And the numbers are still growing".

Another RICS campaign "Property in business" used economist Roger Bootle to raise awareness of the property profession within the business community by, for example, presenting research such as the estimated £18bn each year that is wasted by a failure to manage property assets well. There has also been research into the extent to which business people understand the role of surveyors and the nature of business issues with which they can assist. This campaign generated 5000 enquiries from small businesses on how to contact a surveyor.

The RICS also started the highly successful national newspaper advertorial campaigns profiling those who have recently qualified — an approach that was quickly copied by the other professions. Two such adverts generated 18,000 visits to the RICS careers website — an excellent response by any standard.

RICS came a highly commended second to Toyota's £25m brand revitalisation award in the prestigious Marketing Society's 2003 Awards. Commenting on the brand revitalisation award, Sean laughs: "We have been toying with car industry inspired-slogans for RICS how about: "The profession in front is a chartered surveyor" or "RICS goes from strength to strength". So the favourite hobby of most professions — moaning about the lack of value for money or effectiveness of their professional association — has had its day in property.

Case study — MM Associates' analytical tool for benchmarking a client's media coverage

The process involves press cuttings, covering a client's own press mentions and those of key competitors, being read by MM Associates, which assigns each item a number of categories.

This is then aggregated into statistical data presented as a monthly report containing charts, tables and comment and comparing the performance of the client and a number of its competitors. This includes:

- An overview of monthly performance showing the organisation compared with named competitors.
- Analysis showing the market share of the client organisation and those of its competitors across the national, regional, property and non-property trade press.
- Data on the split between press articles relating to corporate, commercial and residential property business.
- Tables showing the number of press articles for each organisation by publication title.
- Sector analysis, revealing the press profile for specific departments and activities such as investment, retail, offices.
- Research reports that gain media coverage are noted so that the organisation can monitor not only how much coverage individual reports receive but also in which type of publication they appear, ie national, regional or property press.
- Spokespeople are monitored, enabling an organisation to track direct quotes that appear across the press spectrum.

The firm says the benefits of using the service are that it provides: a confidential reporting tool for developing PR strategy; benchmarks an organisation against its competitors; identifies where to focus effort and target resources; delivers an overview of an organisation's press profile; reveals gaps in an organisation's media profile — nationally and regionally; supplies information for measuring the performance of PR agencies; helps in planning press campaigns with a regional or national focus; and motivates individuals to be spokespersons and encourages a focused dialogue with the media.

Useful Information Sources

This is not intended as a comprehensive index for the property sector in the UK. It is designed to provide the contact details for those organisations, outlets and resources that are mentioned in this book. If you think that we have missed any important organisations, please let us know and we will endeavour to include them in future editions. Each entry is listed by source, contact details and comments.

BENNS Media
CMP Information Ltd
Sovereign House
Sovereign Way
Tonbridge
Kent TN9 1RW
Tel: 01732 377591
Website: *www.cmpdata.co.uk/benns/*
A media directory for media professionals providing information on business and consumer press, radio and television. Four-volume set — UK, Europe, North America and World. All four volumes £395 each.

British Rate and Data (BRAD)
EMAP Media
33–39 Bowling Green Lane
London EC1R ODA
Tel: 020 7505 8000
Directory of media — with advertising and circulation/readership information.

British Property Federation (BPF)
7th Floor, 1 Warwick Row
London SE1E 5ER
Tel: 020 7828 0111
E-mail: info@bpf.org.uk
Website: *www.bpf.propertymall.com*
"To sustain and promote the interests of all those who own and invest in property in the UK." BPF National Conference is an important event.

Communications, Advertising and Marketing (CAM) Foundation
The CAM Foundation
Moor Hall
Cookham
Maidenhead, Berkshire SL6 9QH
Tel: 01628 427120
E-mail: info@camfoundation.com
Website: *www.camfoundation.com*
A charity managing the CAM Diploma in Marketing Communications in conjunction with CIM.

Chartered Institute of Journalists (CIoJ)
2 Dock Offices
Surrey Quays Road
London SE16 2XU
Tel: 020 7252 1187
E-mail: memberservices@ioj.co.uk
Website: *www.ioj.co.uk*
The oldest professional body/trade union for journalists. Publishes a directory of freelance journalists with experience in different sectors and subjects.

Chartered Institute of Marketing (CIM)
Moor Hall
Cookham
Maidenhead, Berkshire SL6 9QH
Tel: 01628 427 500
E-mail: info@cim.co.uk
Website: *www.cim.co.uk*
Professional body for marketing people — overseeing their professional qualifications.

Chartered Institute of Public Relations (CIPR)
The Old Trading House
15 Northburgh Street
London
Tel: 020 7253 5151
E-mail: info@cipr.co.uk
Website: *www.ipr.org.uk*
The professional body for the public relations industry. The CIPR diploma is a key qualification for PR professionals. Operates a job shop on the website and a PR services directory. Independent consultant free online search.

Freemans
1st Floor
7 Apple Tree Yard
London SW1Y 6LD
Tel: 0207 484 6666
Website: *www.efreeman.co.uk*
An excellent information source and news service about the UK commercial property industry.

FT Commercial Property Conference
FT Conferences
PO Box 406 West Byfleet
DT14 6WK
Tel: 020 7017 5529
E-mail: info@financialtimesconferences.com
Website: *www.ftconferences.com/property*
A major commercial property conference in September each year — in 2005, the theme was "The Democratisation of Property".

Guild of Professional Estate Agents
121 Park Lane
Mayfair
London W1K 7AG
Tel: 20 7629 4141
Website: *www.property-platform.com*
A network of 700 independent estate agents, offering property marketing economies of scale.

Hollis UK Press & PR Annual
Hollis Publishing Ltd
Harlequin House
7 High Street
Teddington, Middlesex TW11 8EL
Tel: 020 8977 7711
A media directory.

Marketing Resources
75 Gray's Inn Road
London WC1X 8US
Tel: 020 7242 6321
E-mail: gtaylor@marketingresources.co.uk
A marketing recruitment consultancy specialising in recruiting property
and media relations staff for the property industry and the professions.

MediaDisk
See Romeike entry
700,000 contacts and 165,000 media outlets. News distribution service
and clipsource.

Media PocketBook
NTC Publications Ltd
Farm Road
Henley-on-Thames
Oxfordshire RG9 1EJ
Tel: 01491 411000
Media directory.

Media UK
E-mail: admin@mediauk.com
Website: *www.mediauk.com*
Media directory.

MIPIM
Reed MIDEM
Walmar House
296 Regent Street
London W1R 6AB
Tel: 020 7528 0086
Website: *www.mipim.com*

A major annual property exhibition and conference in Cannes, France where 5000 delegates attend. Publishes a daily magazine and organises press conferences during the exhibition and a useful link for journalists and delegates/exhibitors.

MM Associates
Tel: 020 8542 5755
Tel: 07745 454319
E-mail: mary@mmassocs.co.uk
A consultancy specialising in measuring media relations effectiveness in the property industry.

MP Forum
Practice Management International
Warnford Court
29 Throgmorton Street
London EC2N 2AT
Tel: 020 7786 9786
E-mail: mpf@pmint.co.uk
Websites: *www.mpfglobal.com*
www.pmint.co.uk
A networking group for senior people (Managing Partners Forum) and marketers (Marketing Partners Forum) in the professions — including surveyors. Management awards are also included. It publishes *Managing Partner* magazine and *Professional Marketing* magazine. The latter is a good source of educational and news material about marketing in the professions — including property. Pmint website has a job bank to seek jobs/candidates in marketing and media in property.

National Association of Estate Agents (NAEA)
Arbon House
21 Jury Street
Warwick CV34 4EH
Tel: 01926 496800
Website: *www.naea.co.uk*
NAEA has 10,000 members, about one-third of UK estate agents. One of its training courses is "Marketing, Advertising and Information Technology" covering basic marketing but not media relations.

National Union of Journalists (NUJ)
Headland House
308–312 Gray's Inn Road
London WC1X 8DP
Tel: 020 7282 7916
E-mail: info@nuj.org.uk
Website: *www.nuj.org.uk*
A leading union for 35,000 journalists. Operates a freelance directory.

Newspaper Licensing Agency Ltd (NLA)
7–9 Church Road
Wellington Gate
Tunbridge Wells TN1 1NL
Tel: 01892 525273
E-mail: copy@nla.co.uk
Website: *www.nla.co.uk*
If you copy press cuttings for internal circulation, you will need to pay
the appropriate copyright fee.

PRNewswire
209–215 Blackfriars Road
London SE1 8NL
Tel: 020 7490 8111
E-mail: info@prnewswire.co.uk
Website: *www.prnewswire.co.uk*
A media directory and targeting service. Provides electronic
distribution, targeting, measurement, translation and broadcast
services on behalf of 40,000 customers. Services include ProfNet,
eWatch, MEDIAtlas and MultiVu. A subsidiary of United Business
Media plc. Website also provides a headline search.

PRPlanner
See Romeike entry.

PROFILE — the property marketing network
18 Soho Square
London W1D 6QL
Tel: 020 7025 8766
E-mail: info@profile-network.com
Website: *www.profile-network.com*
A networking and educational group for those involved in marketing

in the property sector. Many events involve journalists talking about how to work with their magazine/newspaper.

Property Advertising Marketing And Design Awards (PAMADA)
E-mail: info@pamada.org.uk
Website: *www.pamada.org.uk*
Formerly two separate award events — Estates Gazette Front Cover Awards and Property Advertising Marketing And Design Awards (The Company of Chartered Surveyors), which have now merged. The website lists past winners and their agency details.

Propertymall Information Mall Ltd
36 Great Queen Street
London WC2B 5AA
Tel: 020 7611 5040
E-mail: manager@propertymall.com
Website: *www.propertymall.com*
A key information source for all things to do with commercial property. Reports a variety of news items.

Professional Services Marketing Group (PSMG)
PO Box 131
Saffron Walden
Essex CB11 4ZN
Tel: 020 7642 1179
E-mail: admin@psmg.co.uk
Website: *www.psmg.co.uk*
A professional networking and educational association for those involved in marketing the professions (including property). The website has a news section. PSMG magazine has information about marketing and media relations in the professions. Website has a job bank containing marketing and media jobs in the property sector — for candidates and employers.

Public Relations Consultancy Association (PRCA)
Willow House
Willow Place
London SW1P 1JH
Tel: 020 7233 6026
E-mail: Preview@prca.org.uk
Website: *www.prca.org.uk*

The trade/professional body for public relations agencies. A good source of information when you are selecting PR agencies — a free referral service to clients looking to appoint a PR consultancy.

Royal Institute of British Architects (RIBA)
66 Portland Place
London W1B 1AD
Tel: 020 7580 5533
E-mail: info@inst.riba.org
Website: *www.riba.org*
Professional association for architects. It operates a number of award schemes.

Royal Institution of Chartered Surveyors (RICS)
Royal Institution of Chartered Surveyors Contact Centre
Surveyor Court
Westwood Way
Coventry CV4 8JE
Tel: 0870 333 1600
E-mail: contactrics@rics.org
Website: *www.rics.org*
A number of award schemes. The media office is extremely knowledgeable and helpful. It channels media enquiries to relevant surveyors.

Romeike
Chess House
34 Germain Street
Chesham HP15 1SJ
Tel: 0870 736 0010
E-mail: info@romeike.com
Website: *www.romeike.com*
Provides a range of directories and database products to support marketing and media activities. Mediadisk has 700,000 contacts. PRplanner has 49,000 media contacts. Editors provide media information in six volumes. Willings press guide has three volumes.

Schillings
Royalty House
72–74 Dean Street
London W1D 3TL

Tel: 020 7453 2500
Fax: 020 7453 2600
E-mail: legal@schillings.co.uk
Website: *www.schillings.co.uk*
Weekend emergency number — Tel: 07711 715345
Solicitors who specialise in acting for individuals and organisations in media management, both pre and post-publication.

The Newspaper Society
The Intelligence Unit
Bloomsbury House
74–77 Great Russell Street
London W1B 3DA
Tel: 020 7636 7014
E-mail: ns@newspapersoc.org.uk
Website: *www.newspapersoc.org.uk*
Database of regional and local newspapers in the UK.

UK Press Directory
UK Media Directories Ltd
32 South Road
Saffron Walden
Essex CB11 3DN
Tel: 01799 502665
Media directory.

Willings Press Guide
Chess House
34 Germain Street
Chesham, Bucks HP5 1SJ
Tel: 0870 736 0011
E-mail: enquiries@willingspress.com
Website: *www.willingspress.com*
See Romeike entry
Directory or online — 65,000 publications and media outlets listed. Three volumes — UK, Western Europe and World.

National media contacts

Daily Mail
Northcliffe House
2 Derry Street
London W8 5TT
Tel: 020 7938 6000
Website: *www.dailymail.co.uk*
Property editor: Nigel Lewis
Tabloid property section appearing on Fridays and interested in prices, families and houses plus popular foreign property. Strong emphasis on case studies and has good pictures of properties to accompany stories

Daily Telegraph
1 Canada Square
Canary Wharf
London E14 5DT
Tel: 020 7538 5000
Website: *www.telegraph.co.uk*
Property editor: Angela Pertusini
Broadsheet property section on Saturdays, particularly interested in top-end country properties, international holiday home locations and strong UK regional features, plus gardening.

Financial Times
1 Southwark Bridge
London SE1 9HL
Tel: 020 7873 3000
Website: *www.ft.com*
Property editor: Alison Beard
Broadsheet property pages within Saturday's *Weekend* section, concentrating on design and top-end property in exotic foreign locations. About six special property supplements per year on specific subjects or locations — for example, ski properties or North America.

The Independent
Independent House
191 Marsh Wall
London E14 9RS
Tel: 020 7005 2000
Website: *www.independent.co.uk*

Property editor: Madeleine Lim
Compact property section on Wednesdays featuring design-led architecture and interiors, properties in unusual overseas locations and some investment stories. Section places high emphasis on high-quality pictures and will not use computer images.

Mail on Sunday
Northcliffe House
2 Derry Street
London W8 5TT
Tel: 020 7938 6000
Website: *www.mailonsunday.co.uk*
Property editor: Sebastian O'Kelley
Tabloid property section concentrating on celebrity homes, get-rich-quick property schemes, property market scandals and human interest stories. It often uses celebrity writers from property programmes. This section is circulated only within the Carlton TV area of the Home Counties and London.

The Observer
3–7 Herbal Hill
London EC1R 5EJ
Tel: 020 7278 2332
Website: *www.observer.co.uk*
Property editor: Jill Insley
Property pages carried at the back of compact personal finance section called Cash. It majors on design, environmental aspects of homes, overseas properties and social market issues such as key worker housing.

Sunday Express
Northern & Shell Building
10 Lower Thames Street
London EC4R 6EN
Tel: 0871 434 1010
Website: *www.express.co.uk*
Property editor: Jane Slade
Tabloid property section interested in human interest stories, celebrities, new homes, latest prices.

Sunday Telegraph
1 Canada Square
London E14 5DT
Tel: 020 7538 5000
Website: *www.telegraph.co.uk*
Property editor: Anne Cuthbertson
Broadsheet property section concerned with country properties, the
housing market, investment schemes, gardens and DIY.

Sunday Times
1 Virginia Street
London E1 9BD
Tel: 020 7782 4000
Website: *www.timesonline.co.uk*
Property editor: Carey Scott
Compact property section, widely considered the market leader. It
majors on sales and rental market stories, human endeavour property
projects, high-status overseas homes and gardening.

The Times
1 Pennington Street
London E98 1TT
Tel: 020 7782 5000
Website: *www.timesonline.co.uk*
Property editor: Catherine Riley
Compact property section with a highly informed content concerned
with housing trends, market information, unusual overseas property,
architecture, high-status property and investment issues.

Property and construction media contacts

Architects' Journal
EMAP Business Communications
151 Rosebery Avenue
London EC1R 4GB
Tel: 020 7505 6700
Editor: Isabel Allen
Weekly, carrying features and news about architecture, planning and
buildings.
Target: architects

Architectural Design
John Wiley and Sons Ltd
4th floor
International House
Ealing Broadway Centre
London W5 5DB
Tel: 020 8326 3800
Website: *www.wiley.co.uk*
Editor: Helen Castle
Six editions annually, with features and news on international architecture and design trends.
Target: architects

Architectural Review
EMAP Construct
151 Rosebery Avenue
London EC1R 4GB
Tel: 020 7505 6725
Website: *www.arplus.com*
Editor: Paul Finch
Monthly, carrying features on architecture and design.
Target: architects

Architecture Today
161 Rosebery Avenue
London EC1R 4QX
Tel: 020 7837 0143
Editors: Ian Latham and Mark Swenarton
Published 10 times a year, about modern European architecture.
Target: architects

BBC Good Homes
Woodlands
80 Wood Lane
London W12 0TT
Tel: 020 8433 2391
Website: *www.bbcgoodhomes.co.uk*
Editor: Lisa Allen
Monthly, about property and interiors of mid and lower-range homes, assessed from a popular consumer perspective.
Target: general public

Building
The Builder Group plc
7th Floor
Anchorage House
2 Clove Crescent
London E14 2BE
Tel: 020 7560 4000
Editor: Denise Chevin
Weekly, covering residential, commercial and manufacturing aspects of the building industry, including some overseas stories.
Target: building industry professions and economists.

Building Design
CMP Information Ltd
Ludgate House
245 Blackfriars Road
London SE1 9UY
Tel: 020 7291 8200
Editor: Robert Booth
Weekly, including news and features on all aspects of building design.
Target: building industry professions.

Build It
1 Canada Square
Canary Wharf
London E14 5AP
Tel: 020 7772 8440
Website: *www.self-build.co.uk*
Editor: Catherine Monk
Monthly, about self-building and conversion techniques, including detailed tips and supplier information.
Targets: self-builders and conversion specialists.

Built Environment
Alexander Press
1 The Farthings
Marcham
Oxfordshire
OX13 6QD
Tel: 01865 391518
Editors: Peter Hall and David Banister

Quarterly, featuring articles on architecture, planning and the environment.
Target: architects, planners and academics.

Construction News
151 Rosebery Avenue
London EC1R 4GB
Website: *www.cnplus.co.uk*
Construction industry's leading weekly magazine.
Target: builders, planners and economists.

Country Homes and Interiors
IPC Magazines
King's Reach Tower
Stamford Street
London SE1 9LS
Tel: 020 7261 6451
Editor: Rhoda Perry
Monthly, with features on rural properties, interiors and furniture.
Target: general public

Country Life
IPC Magazines
King's Reach Tower
Stamford Street
London SE1 9LS
Tel: 020 7261 6400
Website: *www.countrylife.co.uk*
Editor: Clive Aslet
Weekly, the quintessential bible of the country set featuring top-end rural property, architecture, interior, gardening and traditional rural sports and lifestyles.
Target: top-end country dwellers

Country Living
National Magazine Co Ltd
National Magazine House
72 Broadwick Street
London W1F 9EP
Tel: 020 7439 5000
Website: *www.countryliving.co.uk*

Editor: Suzy Smith
Monthly, featuring mid to top-end properties, interiors and gardens.
Target: general public

Estate Agency News
Estates Press Ltd
Keenans Mill
Lord Street
St Annes-on-Sea
Lancashire FY8 2ER
Tel: 01253 783206
Website: *www.estateagencynews.co.uk*
Editor: Tony Durkin
Monthly specialist publication for and about estate agents.
Target: residential property professionals

Estates Gazette and EGI
1 Procter Street
London WC1V 6EU
Tel: 020 7911 1700
Websites: *www.egi.co.uk, www.eg.tv*
Editor: Peter Bill
The leading commercial property weekly magazine and online site, with some residential content. High emphasis on sector and regional focus reporting. Press day for EG is Thursday with publication on Saturday. All editorial contacts and the features list appears on the website.
Target: property industry professions

Grand Designs
National House
High Street
Epping
Essex CM16 4BD
Website: *www.granddesignsmagazine.com*
Editor: David Redhead
Monthly tie-in with Channel 4 show on distinctive architecture-driven self-build projects.
Target: self-builders and general public

Homebuilding and Renovating
Homebuilding
Ascent Publishing Ltd
2 Sugar Brook Court
Aston Road
Bromsgrove
Worcs, B60 3EX
Website: *www.homebuilding.co.uk*
Editor: Jason Orme
Articles on building homes from scratch or renovating them from old.
Target: self-builders and renovation enthusiasts

Homes and Gardens
IPC Magazines
King's Reach Tower
Stamford Street
London SE1 9LS
Tel: 020 7261 5000
Editor: Deborah Barker
Monthly, including features on properties and interiors.
Target: general public

House Beautiful
National Magazine Co Ltd
National Magazine House
72 Broadwick Street
London W1F 9EP
Tel: 020 7439 5000
Website: *www.housebeautiful.co.uk*
Editor: Kerryn Harper
Monthly, mainly about property interiors but occasional articles on UK
and overseas property markets.
Target: women

Housebuilder
Byron House
7–9 St James's Street
London SW1A 1DW
Tel: 020 7960 1630
Website: *www.house-builder.co.uk*
Editor: Ben Roskrow

Monthly official journal of the Home Builders' Federation, covering all aspects of the residential property industry.
Target: property professions, economists and academics

Ideal Home
IPC Magazines
King's Reach Tower
Stamford Street
London SE1 9LS
Tel: 020 7261 5000
Editor: Susan Rose
Monthly, concentrating on interiors but some property market pieces.
Target: women

Period Living and Traditional Homes
EMAP East
Mappin House
4 Winsley Street
London W1W 8HF
Tel: 020 7343 8775
Editor: Sharon Parsons
Monthly, featuring articles on renovating homes, plus building and gardening appropriate for period homes.
Target: general public

A Place in the Sun
Brooklands Group Ltd
Westgate
120–128 Station Road
Redhill, Surrey RH1 1ET
Tel: 01737 786820
Editor: Mark Haverstock
Monthly, tie-in with television series and featuring mainstream overseas locations and new developments with some investment advice and "how to buy" guides.
Target: general public

Property Week
Ludgate House
245 Blackfriars Road
London SE1 9UY

Tel: 020 7921 5000
Website: *www.property-week.co.uk*
Editor: Giles Barrie
A leading weekly commercial property magazine. All editorial contact
e-mail addresses and telephone numbers are shown on the site.
Target: property industry professions

Self Build and Design
151 Station Street
Burton on Trent
Staffordshire DE14 1BG
Tel: 01283 742950
Website: *www.selfbuildanddesign.com*
Editor: Ross Stokes
Monthly, featuring news and features on residential self-build projects.
Target: self-builders and architects

What's New in Building?
Ludgate House
245 Blackfriars Road
London SE1 9UY
Tel: 020 7921 8228
Editor: Mark Pennington
Monthly, concentrating on new products suitable for commercial and
residential construction.
Target: building professionals

Broadcast media contacts

BBC News programmes (television and radio)
The News Centre
BBC Television Centre
Wood Lane
London W12 7RJ
Tel: 020 8743 8000
Website: *www.bbc.co.uk*
The BBC operates a bi-media newsroom (that is, radio and television
news bulletins are produced within the same newsroom), so press
releases sent to BBC News should, in theory, be circulated across
television and radio.

BBC TV programmes (all non-news programmes), plus BBC Online and BBC Ceefax

BBC Television Centre
Wood Lane
London W12 7RJ
Tel: 020 8743 8000
Website: *www.bbc.co.uk*
BBC's television documentaries and consumer programmes (such as *Watchdog*) are produced by separate teams not linked to the news division. Press releases should, therefore, be sent to each individual programme editor.

BBC radio programmes (all non-news programmes)
Broadcasting House
Portland Place
London W1A 1AA
Tel: 020 7580 4468
Website: *www.bbc.co.uk*
BBC's radio documentaries and consumer programmes (such as *You And Yours*) are produced by separate teams not linked to the news division. Press releases should, therefore, be sent to each individual programme editor.

Channel 4 programmes (all non-news programmes)
Channel 4
124 Horseferry Road
London SW1P 2TX
Tel: 020 7396 4444
Website: *www.channel4.com*
Channel 4 documentaries or lifestyle property programmes are made by independent production companies (named on the programme end credits) and are not connected with ITN News, which supplies the news to Channel 4.

Channel Five programmes (all non-news programmes)
Five
22 Long Acre
London WC2E 9LY
Tel: 020 7550 5555
Website: *www.five.tv*

Channel Five's documentaries or lifestyle property programmes are made by independent production companies (named on the programme end credits) and are not connected with ITN News, which supplies the news to Channel Five.

GMTV (ITV breakfast programming including news)
The London Television Centre
Upper Ground
London SE1 9TT
Tel: 020 7827 7000
Website: *www.gmtv.co.uk*
GMTV has its own newsgathering operation and press releases should be sent to this section, as well as to ITN, which supplies the rest of the news to ITV.

Independent Television News
200 Gray's Inn Road
London WC1X 8XZ
Tel: 020 7833 3000
Website: *www.itn.co.uk*
ITN is the supplier of news to ITV, Channel 4 and Channel Five; send separate press releases to the ITN newsdesks for each of the three channels.

British Sky Broadcasting (Sky, including Sky News)
6 Centaurs Business Park
Grant Way
Isleworth
TW7 5QD
Tel: 020 7705 3000
Website: *www.sky.com*
All Sky news programming comes from one newsroom, but separate programmes that may cover property angles (eg, Littlejohn) requires separate releases.

Appendix

Glossary

This is not intended as a comprehensive glossary of all journalistic or PR terms but is designed to provide an insight into many of the terms encountered within this book.

Advertorial	Looks similar to an editorial article, but contains text that promotes one product or company. It has been paid for like advertising.
Angle	The particular perspective of the article or story — the main message to be conveyed, the primary emphasis of the story.
Attribution	Where the credit is given to a person for information, a quote or opinion. Anonymous information is unattributed.
B2B	Business-to-business
B2C	Business-to-consumer
Backgrounder	A detailed document providing additional information about an organisation, individual, product or service. Usually accompanies a press release and may appear in a media or press kit or pack.
Blog	Contraction of weblog, an e-mail or online diary, usually written by one person. Not regarded as reliable or authoritative by most journalists.
Brand	"A successful brand is an identifiable product, service, person or place augmented in such a way that the buyer or user perceives relevant, unique added values which match their needs most closely. Furthermore, its success results from being able to

	sustain these added values in the face of competition." (de Chernatory and McDonald)
Brief	The instructions given to a PR consultant or agency. Ideally, it should be a written brief with a clear statement of aims, target audience and budget.
Broadsheets	Describes national newspapers such as the *Sunday Times, Daily Telegraph*, produced in large format and carrying national and international news and business information.
Byline	The name of the journalist or author of an article.
Caption	Copy (or words) describing a photo, image, graphic or chart.
Circulation	The number of copies of the newspaper or magazine that are produced and distributed. It should be audited by ABC (Audit Bureau of Circulation). Not to be confused with readership.
Clippings	Also known as cuttings. A piece of the newspaper or magazine where a particular individual or organisation has been mentioned.
Column inches	A measure of PR success — multiply the length of the piece by the number of columns.
Commission	When a journalist is given a guarantee that a story will be used and paid for by a newspaper, television or radio programme.
Compacts	The name given to those serious-minded newspapers which, since 2000, have changed from broadsheet format to tabloid or near-tabloid size. This includes *The Independent, The Times* and *The Guardian*.
Computer Graphic Image (CGI)	A computerised artists' impression, usually of a new development or property under construction, used for PR or marketing purposes.
Consumer press	Media that are directed at and read by consumers. This is in comparison to trade/technical media, which are read by those in business.
Copy	The written material in a story or press release as opposed to the photographs or other elements of a layout.
Corporate PR	Public relations activities relating to a corporate organisation. As opposed to product PR relating to a particular product.

Corporate social responsibility (CSR)	A business ethics concept that highlights the voluntary role of an organisation in contributing to a better society and environment beyond its financial and capital commitments.
Coverage	The number of people within your marketing campaign that may see your message. Sometimes expressed as "reach".
Cue material	The introduction into a "packaged" television report, read by a newsreader.
Deadline	The date and time by which the news or information must be with the editor in order to be included and gain coverage.
Editor	The person with final responsibility for what does or does not appear in the media.
Editorial	An expression of opinion (or news) using text and images to make an article, feature, story or short piece.
Embargo	A request to journalists to withhold the information (usually in a press release) until a specified time/date. This is discretionary not compulsory.
Exclusive	A news item or story available to one reporter or publication, radio or television station, before being made available to rivals.
E-zine	An online magazine or newsletter.
FAQ	Frequently Asked Questions, usually used on websites.
Feature	An article that goes into depth on a topic, rather than a short news report.
Forward features	A list of planned features. Some newspapers or magazines publish lists in advance and can be requested from an editorial assistant — some are published on websites. They usually contain a synopsis of the piece and a deadline and contact details for the commissioned reporter.
Freebie	A gift or event offered to a journalist by a PR or client in a crude attempt to influence a story's tone.
Freelance writer	A person who is not employed by one newspaper, magazine or broadcast outlet and is instead commissioned by a range of editors to write copy.
Ghost writer	A person who is commissioned to write an article against which another person is bylined.

Hack	A derisory term for a journalist or reporter from the word "hackneyed", which means a word or phrase is so overused that it has lost its meaning.
Headline	The title of a news story, article or press release. It should be factual and convey the key news story.
High resolution pictures	Print publications require professional photographs either in print form, occasionally in transparency form, or increasingly in electronic form at a minimum of 300dpi and 10cm x 10cm in size.
Human interest	Making a potentially dry, dull story more interesting by introducing angles or components involving individuals.
Inverted pyramid	The structure of a news story — which places the most important information at the top and less important details at the end. This allows the editor to easily cut the story if there are space limitations.
Investor relations	A specialist financial PR function for communicating with investors and analysts who are interested in the share price of a company.
Junket	An abusive term for a visit by journalists when transport, accommodation and other facilities are provided by a PR or client. Increasingly frowned upon by editors, who prefer more impartial environments within which stories can be written.
Launch	Usually an event to mark the start of the sale of a development or, in exceptional circumstances, an individual property. Launch events open to the full press are increasingly disregarded by serious journalists.
Leak	When information — often controversial — is passed to the media prematurely. Most leaks are intentional.
Libel	The publication of a statement that exposes the person to hatred, ridicule or contempt; or which causes him or her to be shunned or avoided; or which has a tendency to injure him or her in his or her office, trade or profession or right-thinking members of the public generally.
Lifestyle story	A property story concentrating on the individuals owning or operating it, or the social trends related to the property.

Market story	Mainly or wholly about the state of the property market or related subjects such as property investment or leasing markets.
Marketing	"The management process for anticipating and meeting customer needs profitably." (The Chartered Institute of Marketing)
Marketing mix	Usually meaning the range of activities undertaken by marketing — product, price, place and promotion (PR, advertising, direct marketing, sales promotion and selling).
Mechanical data	Information about how advertisements appear — eg, page size, number of columns and their widths, advertising rates and deadlines. Also information about how artwork should be provided.
Newsgroups	Topic-based, public discussion forum on the internet where messages are posted. Sometimes known as online forums.
Noise	Slang word for the confusion cased by too many messages trying to be delivered at any one time through the media.
Off the record (OTR)	When information is given to a journalist on the understanding that it may be used but will not be attributed to the source. Anything described as "off the record" (OTR) will be respected by the journalist but if a source does not specify it as OTR, it is regarded as "on the record" and may well be used and attributed.
Opportunities to see (OTS)	A calculation showing the number of times a person is likely to see a marketing message. Usually found relating to advertising.
Package	The bundle of information that makes up a television report, consisting of "cue" material read by a newsreader, the reporter's voiceover and pictures giving the substance of the story, then the interview with one or more people, and a concluding "piece to camera".
Piece to camera	When a television reporter looks into the camera and summarises a story.
Pitch	An attempt to sell an idea for a story or feature; between journalists and editors this is increasingly done by a short e-mail.

Portal	A form of "catch all" internet website that contains content drawn from a selection of individual websites (eg, Rightmove, which consists of content drawn from individual estate agents' websites).
Press day	The day that a weekly, monthly or occasional publication goes to the press. Every day is press day for a daily newspaper.
Press pack	Sometimes known as a press kit or media pack. A collection of information and materials — for example, a press release, biographies, photos, FAQs, backgrounders, contact details, brochures etc — assembled into a folder.
Press release	Sometimes called a news release. A short document containing a statement or information released by an organisation to the media.
Press trip	A visit for journalists where transport, accommodation and other facilities are provided by a PR or client. *See* junket.
Property editor	The person in charge of the property pages or property section of a larger publication that deals with a wider range of material.
Public relations	"Public relations is the discipline which looks after reputation, with the aim of earning understanding and support and influencing opinion and behaviour. It is the planned and sustained effort to establish and maintain goodwill and mutual understanding between an organisation and its publics." (Chartered Institute of Public Relations)
Puff	Abusive term for press release or story that crudely and uncritically promotes one product or company with little research or evidence to support its angle.
Pure property story	Mainly about one property, development; also about a geographical location and the properties within it.
Quote	A statement from a person that appears in quotation marks as opposed to other types of information.
Rate card	The published cost of advertising in any media. Often discounts will be available upon request or as a result of multiple entries.
Reach	*See* coverage
Red tops	*See* tabloids

Readership	The number of times a publication is shared with other individuals. Not to be confused with circulation. There may be 5000 copies of a publication (circulation), but if — on average — three people read each copy, then the readership will be 15,000.
Retraction	A correction by the media of information previously reported or published in error.
Sound bite	A short phrase or audio quote taken from a person's interview.
Spiked	When a news story or feature is not used.
Spread	The same story spread over two or more pages.
Sub-editor	The person who checks the accuracy of an article, edits it to fit the available space and writes the headline.
Syndicated articles	This is where the same article appears across a series of publications, sometimes in different countries.
Synopsis	Short outline of the story wanted by a publication, specifying length, angle, tone, possible interviewees and deadline for submission. Usually issued to the commissioned writer, sometimes released to appropriate PRs too.
Tabloid	Technically, a publication half the size of a broadsheet but commonly used to describe a paper that carries popular news expressed in a basic fashion and appealing to a mass audience. Tabloids include *The Sun, Mirror, News of the World, Daily Express* and *Daily Mail*. Some like the *Mirror* and *The Sun* are called "red tops".
Tone	Whether the editorial piece appears positive, negative or neutral.
Trade/technical media	Magazines written for and read by those in trade or specialist areas — as opposed to general consumers.
Universe	The total number of people that read, listen to, watch a type of media.
Video news release	Like a press release, but the message is conveyed by video; sometimes the release can be on a CD-Rom or DVD.
Wire services	Online services used by publications to obtain news from around the world (such as PA News and Reuters) or on particular topics.

Index